Timeless EUROPE

WHITE STAR PUBLISHERS

Contents

TEXT
FRANCO ANDREONE

EDITOR
VALERIA MANFERTO DE FABIANIS

GRAPHIC DESIGN
MARIA CUCCHI

EDITORIAL STAFF
GIADA FRANCIA
ENRICO LAVAGNO
MARCELLO LIBRA

1 THE SNOWY PEAKS OF THE DOLOMITES (LEFT); THE WHITE CLIFFS OF THE CALANQUES (CENTER); THE AUTUMN COLORS OF A FOREST IN GERMANY (RIGHT).

2-3 THE OULANKA JOKI RIVER CROSSES OULANKA NATIONAL PARK.

4-5 JETS OF GLOWING LAVA SPEW FROM THE CRATER OF ETNA.

6 MORNING FOG COVERS THE MOUNTAINS OF THE VOSGES REGION.

7 THE WHITENESS OF THE SHEER CLIFFS OF THE IONIAN ISLANDS.

8-9 THE AURORA BOREALIS BRIGHTENS THE SKY OVER LAPLAND.

10-11 THE KORNATI ARCHIPELAGO IS CHARACTERIZED BY CHARMING INLETS.

Via Candido Sassone, 22/24
13100 Vercelli, Italy
www.whitestar.it

Translation: Amy Christine Ezrin

ISBN-10: 88-544-0157-9
ISBN-13: 978-88-544-0157-0

REPRINTS:
1 2 3 4 5 6 10 09 08 07 06

Printed in China

Preface
Europe

13 WARM SUNLIGHT CARESSES TALL PEAKS COVERED BY SNOW IN THE DOLOMITES.

Europe, with its long history and the role it has played as the cradle of Western civilization, is now one of the world's geographical areas in which an urgent call has gone out to conserve nature. The coexistence of man and nature has been little less than a chronicle of interactions, conflicts, and disagreements. One immediate consequence of continuous human activity is that almost no area in the westernmost part of the Eurasian continent can truly define itself as "virgin," or, in today's terms, "wilderness."

Practically all European environments (even those which have maintained a "natural" look) have been the object of human developmental interest and have been the focus of human activity. Thus, in presenting sites that demonstrate aspects of naturalness or which are of great significance in terms of landscape, we must keep in mind the human background if we are to correctly interpret these areas' current status and future potential.

In Europe, environment-altering interventions in the relationship between man and nature have been occurring since the end of the last Ice Age, 10,000 to 20,000 years ago, and, since the Middle Ages, they have become more widespread and systematic. Most of the original northern forests that once covered a large part of the territory have been wiped out to make way for various types of farms and ever larger cities. At the same time, many swampy areas have been drained. Numerous animal species have fallen victim to these changes, as demonstrated by troubling reductions in their populations. The great predators and herbivores (the former traditionally seen by humans as dangerous competitors or easy prey, and the latter seen as a source of food) were the first to suffer hard-core hunting campaigns, which caused them to disappear from vast areas. By the Middle Ages, the natural habitat of bears had been restricted to partially impassable mountains with protective forest coverage. Today, the brown bear survives only in small isolated sections of the Balkan Peninsula; elsewhere its numbers are minimal, as in the Abruzzo region of Italy and in Austria).

The other great predator, the wolf, has likewise suffered constant persecution over the last centuries, and is now reduced to basic survival numbers in southern Italy, Eastern Europe, and the Balkans. However, it is interesting to note that recently this animal has begun to recover territory and has expanded its habitat from the center of the Apennine Mountains as far as the Alps and even into the Italian borderlands. This has happened mainly because of natural reforestation that has followed the abandonment of tracts of countryside, and because of international efforts to reintroduce species into the territory. The beaver, for example, is currently to be found in some areas of Switzerland and Germany, and the bearded vulture, a threatened avian predator has now been re-acclimated to some areas of the Southern Alps. Zoological authorities have recently launched a project to domesticate and breed many animal species, favoring the introduction and transferal of non-native species to non-traditional habitats. This program has fostered, among other things, a change in the plant and animal balance in numerous ecosystems. In almost no part of the world is it any longer possible to find a 100-percent pure representation of the original fauna or flora within a large environment. Human presence has also strongly influenced European vegetation. Essentially, only Scandinavia and Russia can boast of still relatively natural and unchanged areas. The percentage of original forest-cover surviving today is 2 to 3 percent in Western Europe and up to 5 to 10 percent in Russia. Europe can, however, boast forest-cover equal to about 20 to 25 percent of that on the entire planet, though this is because we can include the vast coniferous forests of Scandinavia and Russia, the chestnut-tree woods of the Caucasus, and the cork-tree groves of the Mediterranean. Significantly, the timber industry has often chosen conifers over the leafy trees of the native forests because of their quick growth rates, and in many areas single plantations of poplars dominate, grown in order to supply pulp for the production of paper. In the poplar groves, the forest floor is constantly sprayed with pesticides; this prevents the existence of invertebrates, while the cultivation of grain crops prevents continuity between the last tracts of woodland. Diversity survives only where there are still "corridors" of vegetation, regardless of how narrow they may be.

Among Europe's most charming places are, of course, the coasts and islands. The shores of the southern regions are bathed by a still relatively clean and transparent sea, though in many zones increasing areas of eutrophic (with seasonl oxygen deficiency) water can be observed. These shores are characterized by the presence of typical Mediterranean brush. Along the coasts that face onto the Atlantic Ocean, on the other hand, vast sandy stretches are prevalent, shaped by powerful tides. It is understandable that many of the most intact natural areas are found on islands, though the fact of their insularity can cause the rapid extinction of species, and is the cause of a general fragility in their ecosystems.

Europe features a series of mountainous chains that make climate and environment markedly different within relatively small areas. Just as were the Himalayas, the Alps were formed as the result of a collision between tectonic plates, in this case the African and Eurasian ones. This "clash" caused the rapid uprising of the land. No less famous than the Alps are the Pyrenees, which divide into three sections: Atlantic (or western), central, and eastern. The Atlantic Pyrenees are the highest, with altitudes that gradually decrease from east to west. The Apennines, stretching 800 miles down the Italian Peninsula are a chain of great importance in term of nature and wildlife, with areas of exceptional environmental importance and diversity. The other mountain chains – the Jura, the Central Massif, and the distant Caucasus and Urals – are also important, and have unique species in their fauna and flora.Finally, waterways

must be mentioned. Nature often finds shelter where water has a close connection to the land, or where fresh water "marries" the sea. The deltas of the major rivers, such as the Danube and the Po, for example, offer great areas of refuge for wildlife. The vast coastal marshes of the Camargue are another example of this, as is the bay of the Somme, in northern France, whose vast spaces are shaped by the slow, constant and immense "breath" of the tides. To introduce Europe's wonderful natural settings (and later volumes will introduce northern hemisphere areas), it will help to record, even in summary form, the most important biomes and the plant and animal populations that characterize Europe's different climatic regions. In general, from north to south, there is a gradual progression from tundra to taiga, from the great leafy forests to Mediterranean scrub, and, even farther south, to desert areas.The tundra is characterized by very low temperatures, minimally fertile soil, and sparse vegetation of lichens, grasses, low and compact bushes resistant to wind and snow. The word "tundra" comes from the Saami (Lapp) word *tunturia*, which means "treeless plains" – which is often what the landscape is. Though there is a superficial thaw during the very short summers, the underlying layer of the ground still remains frozen; hence the term permafrost (technically, a permanently frozen sub-layer of soil). The animals in this biome concentrate their activity in the summer period, when the vegetation reaches its maximum. Above all, moss and lichens represent plant life; trees are non-existent. The animal species have rapid reproductive cycles: they must reproduce before the cold comes and then prepare for the long, bitter winters. The most common mammals living in the tundra are reindeer; the most common bird is the snow grouse, one of the few fowl to live permanently in these places.

Continuing south, the taiga stretches across the sub-Arctic regions of the earth, which are distinguished by the predominance of coniferous forests. It is a humid region, in which summer is longer than in the tundra. On the forest floor, ericaceous (heath family) plants and mosses grow, aided by abundant rainfall and high humidity. The cold continental climate in these areas, with long winters and brief, dry summers, makes the vegetation more similar to that of other biomes, but on the other hand, the taiga is home to many animals, including large mammals such as bears, elk, reindeer, and wolves, and numerous smaller-sized ones like the lynx and the ermine. Among birds, the crossbill and the nutcracker are common; among the amphibians are common frogs and a few species of newts. As for reptiles, there are slowworms, horned snakes, and vipers. In the temperate forest, the woodland areas are composed of leafy plants or deciduous trees – the types that lose their leaves every autumn and grow new foliage every spring. This kind of forest spreads throughout the humid regions of the temperate zone, in which a warm season alternates with a cold winter and in which rainfall occurs throughout the year. During the cold seasons, the forest has a bare appearance, broken by the occasional appearance of a few evergreen species. Then there is the Mediterranean scrub zone, occupying the region facing onto the continent's southern sea. The scrub is composed mainly of plants suited to the hot summer months; they include lamina-leafed or broad-leafed evergreens, coriaceous (leather-like) bushes or small evergreen trees with leaves, or sclerophyll-type (lignified) forest. The climate is mild, with rainy winters and dry summers. The dense vegetation limits the filtration of light through to the ground, thus reducing the number of grassy plants and the amount of growth on the forest floor. Nocturnal animals like foxes, porcupines, and weasels live in the tall, dense scrub. In the daytime it is common to encounter the Hermann tortoise, coluber snakes, and green and common lizards. Migratory and non-migratory birds are numerous, among them blackbirds, robins, and nightingales. Europe has relatively few desert areas. In general, they are modest zones, prevalently sandy, with dunes. Continuing inland, the low shrubs grow taller, giving way to tall scrub where it is quite common to come across isolated stone pines.

15 A STREAM CROSSES THE FOREST FLOOR IN THE BAVARIAN FOREST, IN GERMANY.

16-17 THE IMMENSE BARRIER OF THE MASSIF OF MONT BLANC RISES IN ALL ITS GLORY.

18-19 THE SHORES OF THE MADDALENA ARCHIPELAGO ARE CHARACTERIZED BY AN EMERALD SEA AND CLIFFS ERODED BY THE WINDS.

20-21 A FLOCK OF CRANES FLY OVER THE SNOWY PEAKS OF THE PYRENEES

The Volcanic Island
Iceland

Iceland, indeed the "Land of Ice," is located at 63° N, 161 miles (260 km) from Greenland and 620 from Norway. Ice covers some 15 to 18 percent of the entire territory of this land that presents a unique combination of glaciers and volcanoes, of ice and fire.

Iceland's most famous glacier is the Vatnajökull, which is the largest in Europe. Some 3100 sq. miles (8030 sq. km) in area, it covers about 8 percent of the island and contains approximately 800 cubic miles (3335 cu. km) of ice. Unlike the glaciers of the Alps or other mountain chains farther south in Europe, Vatnajökull does not feature the typical morphology of a tongue of ice channeled through a valley. Its shape is of the "icecap" type: the glacier spreads over top of the mountain, like an immense shield. The winter snows compact and harden, melting and re-freezing as they accumulate. This process is not, however, irreversible; the weight of the snowfalls themselves cause them to shift down the underlying frozen mantle and, with the summer thaw, the part of the glacier that has slid below the limit of the perennial snow melts, reassuming its liquid form.

Vatnajökull's average thickness is between 2000 and 2600 feet (600 and 800 m), with a maximum of 3300 feet (1000 m) deep in the middle. Considered a mountain, the Vatnajökull, rising 6952 feet (2118 m), is Iceland's highest peak. The "cap" of ice hides several active volcanic cones, including the Grímsvötn, which erupted most recently in 1996 and 1998. The Vatnajökull's secondary glacial landscape is an undulating plateau at altitudes ranging between 2000 and 3300 feet (600 and 1000 m), pocked with valleys and gorges. Several tongues of ice and snow radiate out from the glacier and descend down to the sea.

In addition to the magnificent Vatnajökull, Iceland has other great natural sites. One of the most important and fascinating is the protected area of the Fjallabak Reserve. This extends for 116,000 acres (46,950 hectares) over mountainous terrain, sculpted by volcanoes and geothermic activity, covered by lava flows and sand, crossed by rivers, and dotted by lakes – all characteristics that the reserve seeks to preserve in pristine condition. The topography of the volcano within the Fjallabak Reserve is the direct result of the fact that the protected area is Iceland's biggest and most active geothermic zone, second only to Grímsvötn. Overall, tranquility is the Reserve's main characteristic, and it is one that draws thousands of visitors every year.

Another place of exceptional interest is Torfajökull (or Landmannalaugar), where the geothermic activity of the calderas and hot pools reflect Iceland's volcanic history, here enhanced by chromatic variation of minerals in the rocks that ranges from red and yellow to blue and green.

22 AND 22-23 ONE OF THE MOST IMPORTANT PROTECTED AREAS IN ICELAND IS THE FJALLABAK RESERVE, CHARACTERIZED BY MOUNTAINOUS TERRAIN CROSSED BY RIVERS AND DOTTED BY LAKES.

23 TOP MANY SEA BIRDS LIVE IN ICELAND, BENEFITING FROM THE ISLAND'S GREAT TERRITORIAL INTEGRITY AND A SEA FULL OF FISH. ATLANTIC PUFFINS, WHICH NEST ALONG THE COASTS, ARE THE MOST COMMON SPECIES.

- ICELAND - THE VOLCANIC ISLAND - ICELAND - THE VOL

24 TOP AND 24-25 BECAUSE OF HEAVY RAINS AND THE GLACIERS' SUMMER THAW, ICELAND'S RIVERS ALTERNATE BETWEEN LOW WATER IN WINTER AND FLOOD LEVELS IN SUMMER.

24 BOTTOM ICELAND IS A RELATIVELY YOUNG LAND AND SUBJECT TO PERIODIC QUAKES DUE TO VOLCANIC ACTIVITY AND FREQUENT TELLURIC SHIFTS. IT IS NOT SURPRISING THAT GIVEN ERUPTIONS AND AFTERFLOWS, ICELAND'S LANDSCAPE IS TREELESS AND ITS FIELDS ARID AND ERODED.

26-27 THE LANDMANNALAUGAR IS ONE OF THE BIGGEST ATTRACTIONS ON ICELAND: THE PASTEL-COLOR MOUNTAINS, COMPOSED MAINLY OF RHYOLITE ERUPTIVE ROCKS, FORM A LARGE AREA SURROUNDED BY LAVA FLOWS, WITH STREAMS AND POOLS OF HOT WATER.

The Volcanic Island

28 LEFT RESEARCHERS' ANALYSIS OF THE STRATIGRAPHY OF THE LARGE ICELANDIC GLACIERS MAKES IT POSSIBLE FOR THEM TO GATHER MUCH INFORMATION ON THE EVOLUTION OF THIS LAND IN A CONTINUOUS STATE OF CHANGE.

28 TOP AND 29 TOP THE GRÍMSVÖTN VOLCANO IS CONSTANTLY ACTIVE; ITS MOST DESTRUCTIVE ERUPTIONS DATE BACK TO 1996 AND 1998.

28-29 AND 30-31 JÖKULSÁRLÓN GLACIER, LOCATED IN THE SOUTH OF ICELAND, IS THE LARGEST IN EUROPE. HERE, LARGE ICEBERGS CONSTANTLY BREAK OFF AND FLOAT OFF INTO THE OFF-SHORE WATERS, SOMETIMES PRESENTING A THREAT TO SMALL BOATS.

LAPLAND - SWEDEN - LAPLAND - SWEDEN - LAPLAND -

32 TOP ABISKO NATIONAL PARK, LOCATED IN THE TORNE DISTRICT, WAS FOUNDED IN 1909. FAVORED, THANKS TO ITS OROGRAPHICAL LOCATION, BY A PARTICULARLY MILD AND DRY CLIMATE, THE PARK IS HOME TO A WEALTH OF PLANT LIFE AND IS VERY FOR THE PRESENCE OF SOME HIGHLY RARE TREE SPECIES.

32-33 AND 33 LAPLAND IS A LAND FULL OF LAKES FED BY THE WATERS OF NUMEROUS RIVERS, AMONG THEM THE TORNE.

Lapland

Sweden

In the rest of Europe, the Ice Age ended more than 10,000 years ago. This is not the case in Lapland, where it continues to this day. In fact, in this Nordic land, the snow, ice, and low temperatures continue powerfully to affect the lifestyle and culture of the local populations. Lapland does not refer to a state per se, but to a geographical area and, even more important, to a people; it encompasses a large area, essentially the territory stretching north of the Arctic Circle, overlapping Norway, Sweden, Finland, and Russia.

Swedish Lapland is among the best known and most visited Lapp regions, and furthermore, it largely shares the general natural and traditional aspects that characterize the "other Laplands." The landscape is breathtaking. Among the dominant aspects, besides the rightly named tundra, is the series of *fjäll*, a Swedish term indicating treeless hills, similar to England's Lake District "fells." Here the *fjäll* consist of rounded, not very high hills, with few summits exceeding 6000 feet (1830 m). Permafrost and the tundra in the strict sense of the terms are found in few places, but everywhere the flora is clearly of a sub-Arctic type. In Lapland, the tree line is located at a much lower altitude than in southern Europe, at about 2500 to 3000 feet (760 to 915 m). In general, the flora varies in abundance and diversity according to the soil. Whereas on acidic and gravelly soils the plant species are few and little differentiated, in places where silt was deposited, the flora is particularly lush. Lichens are very abundant and represent the primary source of food for reindeer. One of the most important aspects of the fauna, both for its composition and habits, is the big herds of this cervid. For thousands of years, the reindeer has been a source of wealth and sustenance for the sub-Artic populations, which use it in a variety of ways, depending on the local tradition, at times as a pack animal, or for meat, milk, and skins, or even to be ridden.

The bird life of Lapland is very abundant during summer, whereas there are few large-size mammals other than the reindeer. Brown bears and wolverines are the main predators, whereas the elk come as far as the tree line. As is often the case in northern European areas, insects proliferate in the marshy zones. Both science and anecdote confirm that during the summer, mosquitoes and sand flies are a terrible nuisance to human beings.

In Lapland, the weather is very variable; it may seem to offer three different seasons within a single day. The Lapp summer is as hot and intense as winter is cold and long. When the sun shines, it does so for a hundred days and nights. Experiencing the midnight sun is a singular experience!

34 TOP AND 35 A LARGE PART OF THE ICELANDIC TERRITORY IS OCCUPIED BY CONIFEROUS FORESTS AND BIRCH-TREE WOODS, IN WHICH NUMEROUS ANIMAL SPECIES COEXIST.

34 BOTTOM A YOUNG ELK WANDERS THE CONIFEROUS FORESTS OF THE TAIGA. THIS LARGE HOOFED CREATURE OF THE GREAT NORTH HAS FOUND THE SPACE IT NEEDS TO SURVIVE IN LAPLAND.

Lapland

36 TOP ONE OF THE DOMINANT ASPECTS OF LAPLAND, BESIDES THE TUNDRA, IS THE SERIES OF FJÄLLS, ROUNDED, NOT VERY TALL MOUNTAINS, WITH FEW PEAKS EXCEEDING 6500 FEET (1980 M).

36-37 THE DELTA OF THE RAPA RIVER IN SAREK NATIONAL PARK IS A LARGE PROTECTED AREA SURROUNDED BY TALL MOUNTAINS COVERED YEAR-ROUND BY GLACIERS.

37 TOP THE GOLDEN EAGLE, THE TRUE KING OF BIRDS, IS VERY COMMON IN LAPLAND AS IT PREFERS TO LIVE IN VAST SPACES, ABOVE ALL MOUNTAINOUS.

37 CENTER IT IS IMPOSSIBLE TO THINK OF LAPLAND WITHOUT THINKING OF THE REINDEER. THESE SEMI-DOMESTIC ANIMALS ALTER THEIR EATING HABITS AND CHANGE PASTURELANDS OVER THE COURSE OF THE YEAR: IN SPRING AND SUMMER, THEY RANGE THE MARSHLANDS WHERE THEY FEED ON AQUATIC AND SWAMP PLANTS, WHEREAS IN WINTER THEY FEED ON LICHENS.

37 BOTTOM IN AUTUMN AND WINTER THE TUNDRA'S SUBSOIL REMAINS CONTINUOUSLY FROZEN.

38-39 THE AURORA BOREALIS IS AN OPTICAL PHENOMENON CHARACTERIZED BY THE APPEARANCE OF LIGHT IN THE NIGHTTIME SKY, AT TIMES ACCOMPANIED BY A SOFT SOUND.

39 THE APPEARANCE OF THE AURORA BOREALIS IS DUE TO THE VISIBLE EFFECT OF THE SOLAR WIND THAT, TRAPPED BY THE EARTH'S MAGNETIC FIELD, BUILDS UP. WHEN THE ACCUMULATED CHARGE BECOMES EXCESSIVE, THE CHARGED PARTICLES DISCHARGE IN THE DIRECTION OF THE EARTH'S ATMOSPHERE, PRODUCING THE LIGHT EFFECTS.

The Region of the Thousand Lakes
Finland

Finland can boast an exceptional natural topography: marshes, lakes, and ponds occupy about a third of its territory, a much higher proportion than in any other place in Europe! However, a good part of Finland's marshy areas are covered by forests, and about half the tree-clad area has been drained in order to facilitate tree-felling and timber-hauling. The abundance of the wetlands is also reflected in the botanical species, largely those common to marshlands.

A census of the Finnish lakes has indicated an approximate count of 190,000! It is not surprising, therefore, that some areas in Finland are so full of reservoirs as to each deserve the nickname "The Land of Lakes." This is the case in the Mikkeli region, home to the big Lake Saimaa, a true paradise for anyone in search of peace and quiet: on the shores and in the cottages around the edge of the lake, only the murmur of the wind and the slow movement of the waves can be heard. Saimaa, one of the five biggest lakes in Europe, encompasses 18 lake basins surrounding more than 1000 islands. In the waters of Lake Saimaa lives an endemic subspecies of seal, one of the most threatened creatures in the world. Until recently almost extinct, the seal was once counted in thousands. This typically marine mammal's presence can be explained by the fact that in the past Lake Saimaa was connected to the Baltic Sea. Thereafter, when lower water levels closed the outlet to the sea, a unique ecosystem developed, in which this seal became differentiated from the marine seal.

By the end of the 19th century, pollution had almost caused the extinction of this survivor species. Now, deservedly protected, its population has begun to increase. Nonetheless, the warming winter climate over the last few years has had a harmful effect on current numbers: lack of snow prevents the few females from burrowing into snowbanks and building protective dens, and their young, born without shelter, are at the mercy of predators. Thanks to recently adopted far-sighted protective measures, breeding conditions and safety are likely to improve, raising hopes for the survival of the baby seals. But with a current population of only 250 individuals, this mammal (one of very few lake seal species) is still considered one of the most highly threatened species in Europe.

40 AND 40-41 FROM THE PEAK OF MOUNT KOLI ONE CAN ADMIRE LAKE PIELINEN, A BASIN THAT IS AN INTEGRAL PART OF THE VUOKSI RIVER, LONG AN IMPORTANT WATERWAY FOR THE TRANSPORT OF TIMBER.

41 TOP THE FAUNA OF FINLAND IS VERY ABUNDANT AND DIVERSE. AMONG THE LARGE MAMMALS FOUND THERE, THE BROWN BEAR STANDS OUT; IT HAS ADAPTED WELL TO THE CLIMATE OF THIS REGION.

42 TOP AND 42-43 THE BASIN OF THE SAIMAA LAKE, WHICH OCCUPIES A LARGE PART OF THE SOUTHERN PORTION OF EASTERN FINLAND, IS COMPOSED OF A MAZE OF LITTLE ISLANDS OF DIFFERENT SIZES AND SHAPES.

43 THE ORIGIN OF SAIMAA LAKE DATES BACK TO THE LAST ICE AGE, WHEN THE THICK MANTLE OF ICE THAT COVERS FINLAND COMPLETELY THAWED DUE TO THE RISE IN TEMPERATURE.

Thousand Lakes

44-45 THE FINNISH LAKES ARE VERY IMPORTANT FOR THE ECONOMY, INASMUCH AS THEY ARE USED TO TRANSPORT TIMBER, WHICH COMES FROM THE REGION'S NUMEROUS FORESTS.

45 THE ABUNDANCE OF WETLANDS IN FINLAND IS REFLECTED IN ITS BOTANICAL SPECIES, LARGELY TYPICAL OF MARSHLANDS.

Oulanka National Park

Finland

Oulanka National Park, located in Kuusamo and Salla counties in a mountainous region in northeastern Finland, offers an extraordinarily rich array of elements of the Nordic natural environment. Extending over 88,000 acres (35,600 hectares), the park is unique among its kind for its landscapes as much as for its vegetation. It is probably the protected area in Finland best known abroad.

The Oulanka River, which gives the surrounding area its name, crosses the park together with its tributaries, forming valleys with gorges and sheer rock faces where roiling rapids offer truly fascinating views. Many place names in the park attest to an ancient Lapp presence. The word Oulanka, for example, is of Lapp origin and means "flood" and, in perfect agreement with its name, the river overflows it banks every spring and the rapids become fiercer than ever. The park's geographical location is one of the fundamental reasons for its vast botanical abundance – it has some 500 vascular plant species, i.e., those species with with tissues that conduct water. The soil substratum, rich in nutrients, is a prerequisite for a flora as demanding as that found in Oulanka, where the topography and various microclimates create conditions for the co-existence of a large number of species in the same place. In fact, a great variety of plant interaction can be observed in the park: from fir-tree forests growing on ground covered by thick mosses to sandy and arid terrain with pine trees; from valleys with rocky walls and hills covered by woods to mixed forests and swampy areas of all kinds. In the park, more than 500 plant species, many of which are very rare, have been found.

Among the most beautiful, there are, for example, rare orchids like the *Epipactis atrorubens*, the *Calypso bulbosa*, known as the fairy slipper, and the yellow-toned lady's slipper. A vast wealth of species also characterizes the park's fauna. The riverbeds and flood plains are habitats chosen by many rare butterflies and a hundred or so species of nesting birds. In Oulanka, many species of northern and southern origin and affinities live and grow up together with a certain number of eastern species, whose westernmost limits of distribution run through the park region. Furthermore, the Oulanka River's valley was – after the last Ice Age – an important route for expansion into Finland for species of eastern origin.

46 IN AUTUMN, THE TREES TAKE ON A THOUSAND HUES, CREATING THE SHADE CALLED RUSKA IN FINNISH.

46-47 RIVERS, ABUNDANT AND SWELLING WITH WATER, ACCOMPANY VISITORS THROUGH CONIFER AND BIRCH FORESTS.

47 TOP REINDEER ARE VERY IMPORTANT IN THE FINNISH ECONOMY.

48-49 OULANKA IS CHARACTERIZED BY VAST FORESTS OF FIR TREES AND PEAT BOGS, WITH PONDS, LAKES, GORGES, AND WATERFALLS. MANY PLACES ALSO ATTEST TO THE ANCIENT SAAMI CULTURE: THE NAME OULANKA MEANS "FLOOD," REFLECTING THE REGULARITY OF FLOODING HERE.

50-51 ELK, TOGETHER WITH REINDEER, ARE THE HOOFED CREATURES THAT TAKE THE MOST ADVANTAGE OF THE ABUNDANCE OF FOOD AND BOUNDLESS SPACES.

51 THE TAIGA CLIMATE OF OULANKA INCLUDES SEVERAL ECOSYSTEMS. CHARACTERISTIC OF THE AREA ARE PEAT BOGS, OF PARTICULAR IMPORTANCE TO A LARGE NUMBER OF ANIMAL SPECIES, AMONG THEM MANY AQUATIC BIRDS.

52-53 OULANKA PARK IS UNIQUE IN TERMS OF ITS VARIOUS LANDSCAPES, SOIL VARIETIES, AND PLANT LIFE. IN WINTER, ALL LIFE FREEZES AND THE CONIFEROUS FORESTS TAKE ON A FAIRY-TALE APPEARANCE.

Oulanka

54 TOP AND 54-55 CHARACTERIZED BY HARSH PEAKS AND DEEP FJORDS, THE LOFOTEN ISLANDS RISE HUNDREDS OF FEET STRAIGHT UP OUT OF THE SEA. AT THEIR BASE, CLINGING TO THE REMAINING STRIPS OF GREEN LAND, LITTLE MULTICOLORED HOUSES ARE HOME TO A TENACIOUS COMMUNITY OF FISHERMEN.

55 BOTTOM BIRDS BENEFIT MORE THAN ANYONE FROM THE NATURAL ENVIRONMENT OF THE LOFOTENS, WITH ABOUT 100 SPECIES CATALOGUED SO FAR. THERE ARE NOT ONLY BIRDS OF PREY BUT ALSO SEA BIRDS, LIKE THE BLACK-LEGGED KITTIWAKES THAT REGULARLY NEST IN THE TALL CLIFFS OF THE ARCHIPELAGO.

The Lofoten Islands

Norway

S - NORWAY - THE LOFOTEN ISLANDS - NORWAY - THE LO

Cod fishing is the main occupation in the Lofotens, a group of islands located at 70° N, very close to the Arctic Circle (66° 30' N). The islands are part of the kingdom of Norway. Their surface area is 2000 sq. miles (5200 sq. km), with a total coastal length of at least 2480 miles (6425 km). The inhabitants number about 80,000, though their number varies depending on the season, doubling during the great cod-fishing period, which runs from February to April.

Forged from granite and gneiss, raised during the Caledonian-period wrinkling of the earth's crust (about 400 million years ago), the Lofotens look like alpine spires literally "cast" into the sea. Nonetheless, despite the northern latitude and the archipelago's predominantly barren appearance, the islands enjoy an overall mild summer, similar to late spring in the countries of Mediterranean Europe. Winter, though severe, is never too prohibitive. The Gulf Stream mitigates the excessive cold, displacing the ice-bearing waters. The great glaciers that cover some of the islands descend as far as the sea, digging out wide glacial valleys. Some areas, however, are never covered by glaciers and for this reason host a relict flora (living remnants of extinct species) with a vast variety of living species.

For the Lofoten islanders, the sea is undoubtedly their primary source of income; the entire local economy has always been based on catching and processing cod. Given their strategic position between the cold Arctic waters and the warmers ones to the south, the Lofotens are an obligatory point of passage for many ocean fish, which explains why fishing has always been particularly lucrative in these waters.

The islands' fauna is highly abundant and diverse. Any nature enthusiast who comes to the archipelago will be amazed by the clamor produced by the great colonies of seabirds, which, as often happens in northern archipelagos, find their preferred habitat among the big, sheer cliffs. Among the most widespread seabirds nesting on the Lofotens are Atlantic puffins, black-legged kittiwakes, gannets, and fulmar petrels. Unfortunately, however, some bird species have disappeared because of man. The great auk, for example, similar to a southern-hemisphere penguin, became extinct in the northern Atlantic during the 19th century.

In the past, large-scale plundering of the eggs of nearly all the seabird species occurred. Some of the egg harvest was consumed fresh, some preserved in dry sand for consumption during the winter. Today egg-taking is strictly regulated, with limits that ensure that removing eggs will not affect the biological cycle and the population numbers of a species.

The Lofoten Islands once had a very large population of bears, but these were hunted to extinction. On the other hand, another large mammal, the elk, survives in some areas on the bigger islands. Among sea mammals, the seal and at least six species of whales should be noted; they find bays full of millions of small crustaceans, including krill, their main source of nutriment.

56 THE LOFOTENS ARE PARTICULARLY MAGICAL IN WINTER, A SEASON THAT TRANSFORMS THEM INTO A TRUE PARADISE FOR NATURE LOVERS AND THOSE WHO APPRECIATE THE CHARM OF LONG ARCTIC NIGHTS.

56-57 THE PACE OF LIFE ON THE LOFOTENS IS MARKED BY THE RHYTHMS OF COD FISHING, WHICH TOGETHER WITH TOURISM CONSTITUTES THE MAIN SOURCE OF INCOME FOR THE ARCHIPELAGO.

The Lofoten Islands

The Lofoten Islands

58 TOP THE WATERS OFF THE LOFOTENS, THANKS TO THE GULF STREAM, NEVER FREEZE DURING WINTER. FOR THIS REASON, THEY ARE HOME TO SEVERAL FISH SPECIES, WHICH BETWEEN OCTOBER AND DECEMBER ATTRACT NUMEROUS SCHOOLS OF KILLER AND OTHER WHALES.

58-59 AND 59 BORN OF THE GLACIERS THAT ERODED THE MOUNTAINS, THUS BREAKING THROUGH TO THE SEA, THE NORWEGIAN FJORDS ARE AMONG THE EARTH'S MOST IMPOSING NATURAL FORMATIONS.

60 TOP LEFT THE SPRING PASQUE FLOWER (PULSATILLA VERNALIS) IS THE EMBLEM FLOWER OF DOVREFJELL PARK.

60 TOP RIGHT TORRENTIAL RIVERS, WHICH OFTEN FORM WATERFALLS, BRING TO MIND THE PRIMORDIAL FORCES OF NATURE.

60-61 AND 61 BOTTOM DOVREFJELL IS THE ONLY TUNDRA ZONE IN NORWAY TO NOT HAVE BEEN ALTERED BY HUMAN ACTIVITY; ITS LARGE MARSHY VALLEYS AND VAST BIRCH FORESTS REMAIN UNSPOILED.

Dovrefjell National Park

Norway

The true heart of the mountainous Dovrefjell National Park in central-southern Norway can best be seen in all its glory and diversity in the area between Sunndalsfjella and Knutshøene. Here the landscapes of the Nordic west give way to softer and rounder shapes, as can be seen while crossing the park. Its most admired peaks are the Storkalken in the west, the Storskrymten, and the famous Snøhetta, farther east.

The natural setting reveals great geological diversity, with bare crystalline rocks in the western part and calcareous rocks to the east. These latter favor the growth of a generally abundant plant coverage. Among the park's most interesting features are the Knutshøene hills and the Åmotan Falls, created by the union of five rivers.

The Dovrefjell is currently divided into two distinct parks, the Dovre, created in 2003, and the Dovrefjell-Sunndalsfjella. The region also boasts a northern mountainous ecosystem unique to Norway, still relatively intact and little affected by human activity. The Dovrefjell-Sunndalsfjella National Park encompasses environments that vary from rich wetlands and lush peat bogs to bare hills, but it is also home to many rare plant species that survived the Ice Age. It is no surprise that for over 200 years, passionate environmentalists and botanists have come to this area to admire the great abundance of the flora and fauna.

Unfortunately, even here the activities of modern man have disturbed the mountain ecosystem's original equilibrium, though some remnants of an unspoiled mountain environment are intact in the Dovrefjell-Sunndalsfjella region.

The area is home to many typically northern species, among them the reindeer, the wolverine, the Arctic wolf, and the golden eagle, but the "guests of honor" are without doubt the musk oxen, unique to Dovrefjell. These large herbivores, now found in only a few places in the world, were introduced to the Dovrefjell zone 40 years ago. They were brought from Icelend (their natural habitat), and have been able to form significant herds; the last census recorded 108 musk oxen in the park. In addition, Dovrefjell has about 4500 reindeer – an animal that only 100 years ago seemed threatened with extinction because of intensive hunting.

That crisis has passed, but in more recent times roads, railways, and electricity plants, along with other disturbing factors, threaten these animals, severely hindering their annual migration to locations of scarce mountain vegetation. If denied their migratory routes, reindeer are at risk of losing their ability to survive.

Dovrefjell

62-63 AND 63 TOP IN DOVREFJELL, THE UNCONTAMINATED NATURAL ENVIRONMENTS ARE OFTEN ACCOMPANIED BY AN ABUNDANT AND HIGHLY DIVERSE FAUNA, INCLUDING REINDEER, WOLVERINES, ARCTIC WOLVES, EAGLES, AND MUSK OXEN, WHICH HAVE ACCLIMATED TO THE REGION AFTER BEING INTRODUCED AT THE BEGINNING OF THE TWENTIETH CENTURY.

63 BOTTOM ONE OF THE BIRDS MOST CHARACTERISTIC OF DOVREFJELL IS THE SNOW GROUSE, WHOSE SUMMER PLUMAGE IS A BROWN COLOR WITH GREY SPOTS, THE ODD WHITE FEATHER, AND WHITE WINGS AND CHEST.

64 TOP AND 64-65 THE CAIRNGORM HIGHLANDS FEATURE THE UNITED KINGDOM'S LARGEST AREA OF TUNDRA.

65 THE PARK IS HOME TO A FOURTH OF THE SCOTLAND'S WOODLANDS AND PROVIDES SHELTER TO MANY RARE PLANTS AND ANIMALS. THE RIVERS THAT RUN THROUGH THE PARK, THANKS TO THEIR CRYSTALLINE WATERS, ARE FULL OF SALMON AND ARE ALSO USED TO MAKE EXCELLENT WHISKIES.

Cairngorms National Park

United Kingdom

CAIRNGORMS NATIONAL PARK - UNITED KINGDOM - CAIRNGORMS

The peaks of the Cairngorm Mountains in east-central Scotland provide a true taste within Europe of the Arctic environment. In fact, the Cairngorm region is home to the United Kingdom's biggest mountainous massif, inside which are hidden some of Scotland's most torrential rivers. These mountains present very different features than do other regions of Scotland: they feature vast expanses of granite with moraines, deep-set valleys, and lakes in a pattern seldom found elsewhere, though similar landscapes can be seen in Arctic Canada's Baffin Island

The Cairngorms' tallest peak, Ben Macdui (4295 ft/1310 m), leads several others that exceed 3000 ft (915 m), among them Braeriach (4249 ft/1295 m), Cairn Toul (4127 ft/1258 m), and Cairn Gorm (4085 ft/1245 m). Glacial erosion has leveled most the Cairngorm peaks and has formed some plateaus, often featuring deep depressions (the Avon and Dee lochs). Because of the particular microclimate, snow lasts a long time on the Cairngorms and animals and plants have had to adapt to surviving in such extreme conditions. Many local plant, bird, and insect species are typical of Arctic environments, thus the Cairngorm area, one of the last wilderness zones in Western Europe and, in Scotland, is often called "Arctic Britain."

As it crumbles away, the granitic rock characterizing the Cairngorm massif generates soils of limited depth that are generally rather infertile. Because of this, no large settlements have ever arisen in the area; it has proved to be impossible to cultivate the land or develop traditional sheep-raising farms.

On some hillsides at low altitudes, tracts of Britain's ancient coniferous forest survive. In the past, a large part of Scotland was forested, and the surviving tracts, mainly in the west and known as the Caledonian forest, make it possible even today to experience the atmosphere of the great woodland areas of northern Europe. These hill-and-forest zones are an important habitat for woodland fauna, among them red squirrels, wild cats, and martens. The golden eagle regularly nests in old trees, also on the cliffs that rise sheer above valleys and peat bogs. Deep in the remnants of the ancient forests are some of the most characteristic Scottish lochs, fed by remarkably pure rivers that contribute powerfully to local natural environments.

66 TOP AND 66-67 THE PARK, ABOVE ALL IN AUTUMN, IS CLOAKED IN THE COLORS AND ATMOSPHERE TYPICAL OF NORDIC LANDSCAPES, WITH BOUNDLESS MOORS AND MISTS THAT WIPE THE LANDSCAPE AWAY.

67 TOP SIKA DEER ARE VERY ABUNDANT IN THE OUTLYING AREAS OF THE CAIRNGORMS. THESE DEER ARE A SPECIES INTRODUCED BY MAN AND THEIR EXPANSION IS THE CAUSE OF SOME WORRY, AS THE ANIMAL BREEDS EASILY WITH RED DEER, THUS COMPROMISING THEIR GENETIC INTEGRITY.

67 BOTTOM THE PEAT BOGS FOUND IN THE AREA OF THE CAIRNGORMS FORM INTERESTING AQUATIC ECOSYSTEMS, SOME OF WHICH HAVE RETURNED TO LEVELS OF REMARKABLE NATURALNESS.

68-69 LOCATED AT THE BASE OF THE CAIRNGORMS, LOCH MORLICH IS A VERY POPULAR AREA WHERE SEVERAL WATER SPORTS CAN BE ENJOYED.

70 TOP THE PARK IS CHARACTERIZED BY ALTERNATING AREAS OF MOORS, WOODS, AND GREEN VALLEYS.

70-71 LAKE WINDERMERE, LOCATED IN THE SOUTHERN PART OF THE PARK, IS ABOUT 10 MILES (16 KM) LONG.

71 LAKE THIRLMERE WAS FORMED BY THE CONSTRUCTION OF A DAM.

72-73 THE TOWN OF KESWICK STANDS ON THE SHORES OF DERWENTWATER.

THE LAKE DISTRICT NATIONAL PARK - UNITED KINGDOM

The Lake District National Park

United Kingdom

England's Lake District, which is a perfect picture postcard in summer and autumn, with the bare profiles of the mountains contrasting with the red of the oak forests and the first bands of the snowline, is certainly among the best loved and most popular attractions for British as well as for foreign visitors. Undeniably, the majestic landscapes were a key factor in the establishment of this region of northwestern England as a national park. It is noteworthy that at the time of the Lake District National Park's foundation, Beatrix Potter, a local resident and author of the famed children's book series, donated extensive farmland to the National Trust, thus promoting the development of the park.

The Lake District National Park is Britain's largest: every year it records some 12 million visitors, enchanted by the many lakes (there are 16 main ones). As can be deduced from the shapes of their basins, the lakes were formed over the course of the Ice Age, when great glaciers carved out the beds of stone by force, depositing vast end moraines.

The Lake District is a region of spectacular contrasts: it is above all in the combination of the green of the meadows, the blue of the lakes, and the grey of the mountains that a difficult-to-describe harmony is created. However, thousands of years of human activity have altered the appearance of the landscapes. Lakes, farms, woods, and pasturelands give every valley its own particular appearance and special culture. The semi-natural woods add spectacular patterns, colors, and variety to the landscapes, not to mention shelter to the native animals and plants. The high levels of rainfall in the heart of the national park foster the growth of woods abundant in Atlantic mosses, ferns, and lichens.

The Lake District is home to England's highest peaks (only 7 exceed 3000 ft/915 m) and, consequently, to the territory of a wide variety of birds characteristic of mountain ecosystems. In spring, many migratory species arrive, while throughout the entire year, nesting species can be spotted, among them crows and carrion crows, buzzards, peregrine falcons, and golden eagles. At lower elevations, the environment features more numerous trees, with extensive forests of oaks. Among the many migratory species that can be encountered in spring, the spotted flycatchers (*Muscicapa striata*) that come to Europe from Africa and India, are particularly interesting. Recently, ospreys have returned to the Lake District and can be observed both in spring as well as in autumn, when they migrate between their breeding grounds and their winter territory.

74 TOP THE LAKE DISTRICT NATIONAL PARK IS LOCATED IN CUMBRIA COUNTY, BETWEEN PRESTON AND CARLISLE AND BETWEEN CUMBRIAN MOUNTAINS AND THE IRISH SEA.

74 BOTTOM AND 74-75 DERWENTWATER IS FED BY THE RIVER DERWENT, WHICH CROSSES THE GREEN BORROWDALE VALLEY. THE LAKE HAS FOUR SMALL ISLANDS, NOW OWNED BY THE NATIONAL TRUST.

76-77 LAKE THIRLMERE, ONLY RECENTLY ACCESSIBLE TO VISITORS, IS SURROUNDED BY DENSE WOODS OF LARCH AND FIR TREES THAT COVER HELVELLYN HILL'S SOFT SLOPES.

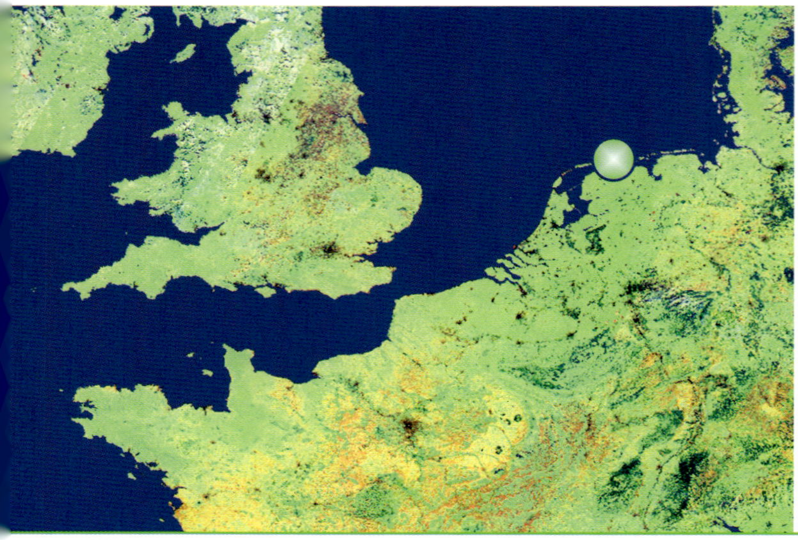

THE WADDEN ISLANDS - THE NETHERLANDS - THE WAD

82 TOP AND 82-83 IN THE HIGHEST AND DRIEST ZONES OF TEXEL ISLAND, HEATH-TYPE VEGETATION GROWS, COMPOSED OF ERICACEAE. NONETHELESS, MANY AREAS ARE CHARACTERIZED BY THE PRESENCE OF VAST SWAMPS AND MARSHES, WHICH FAVOR THE GROWTH OF SEDGES AND RUSHES.

83 THE WADDEN ISLANDS ARE CROSSED BY NUMEROUS CANALS, WHICH LINK THE COAST TO THE SAND DUNES, THE PASTURELANDS TO THE FARMLANDS, AND THE WETLANDS TO THE GREEN FIELDS.

The Wadden Islands
The Netherlands

Texel Island is the southernmost and the largest of the West Frisian Islands, in the Waddenzee, about 18 miles (28,96 km) west of the Netherlands coast; since 1986 it has been a national park. Its particular claim to fame is its abundance of birdlife: every year, about 80 species nest here.

The West Frisian Islands are separated from the mainland by the Waddenzee, a tract of shallow water that often dries up completely at low tide. The islands are young formations, dating back only 8000 to 13,000 years. Around 10,000 years ago, the level of the sea was about 200 feet (60 m) lower than it is today. At that time, the great Ice Ages of the Quaternary period had just come to an end, thus a good part of the water was trapped in the big glaciers that still covered Scandinavia and Scotland. The North Sea had not yet formed, and a few big rivers, among them the Thames, the Elbe, the Meuse, and Rhine, flowed into this area. The area that separated Great Britain from the continent consisted of a wide, fresh-water-filled basin. The subsequent climatic warming made the sea level rise; the waters of the Atlantic then rushed into this basin so that sand and sediment amassed along the coast to finally form a long series of dunes. The sea thus isolated a group of islands and invaded the depression that separated them from the continent, forming the Waddenzee, a small basin with a silt floor. Just before the year 1000, the local populations began to consolidate the terrain, and a few centuries later, they began to erect dykes to protect themselves from storms and floods. Thus, the lands reclaimed from the sea were gradually drained and turned into farmland, a huge project that was finished around the 17th century, when Texel Island took on its present-day appearance.

The long pattern of the tides is at the heart of the West Frisian Islands' biological rhythms: the tide's alternations, in fact, temporarily reveal vast sand and mud banks, loaded with prey (small invertebrates) and nutritious substances representing an irresistible call to birds and other organisms. Whereas oystercatchers are everywhere, as are herring gulls, redshanks, and godwits, the true "star" of the islands' bird life is the spoonbill, the symbol of the new national park.

Marine mammals also find refuge on Texel. Among these, the grey seal and the common seal take advantage of the abundance of prey in the Waddenzee. Because of hunting and pollution, the number of these pinnipeds had fallen to fewer than 500 in 1975. Since then, the population has increased markedly, thanks to seals becoming a protected species. In addition, there has been an immigration of seals from Germany. Unfortunately, in 1988, a highly infectious disease spread among the seals and again reduced the population. Since then, the colony has slowly rebuilt itself; it currently contains 11,000 common seals and a few thousand grey seals.

The Wadden Islands

84 TOP THE BIRDLIFE IS HIGHLY INTERESTING ON TEXEL AND ALL THE WADDEN ISLANDS. FROM THE LITTLE STREETS THAT FLANK THE POLDER (AREAS OF LAND RECLAIMED FROM THE SEA BY DYKES) IT IS POSSIBLE TO OBSERVE SEVERAL SPECIES OF BIRDS, AMONG THEM CURLEWS, REDSHANKS, OYSTER CATCHERS, KESTRELS, BUZZARDS, BRANT GEESE, WILD GEESE, AND BLACK-LEGGED KITTIWAKES.

84-85 ARTEMISIA MARITIMA, SHOWN HERE IN BLOOM IN TEXEL, GROWS MAINLY IN AREAS PROTECTED FROM THE NESTING OF WILD GEESE.

85 TEXEL IS KNOWN AS THE "ISLAND OF THE BIRDS": BIRD SPECIES ARE, IN FACT, NUMEROUS, WITH THOUSANDS OF GEESE, WADERS, LIMICOLOUS SPECIES, HERONS, BIRDS OF PREY, AND SPOONBILLS.

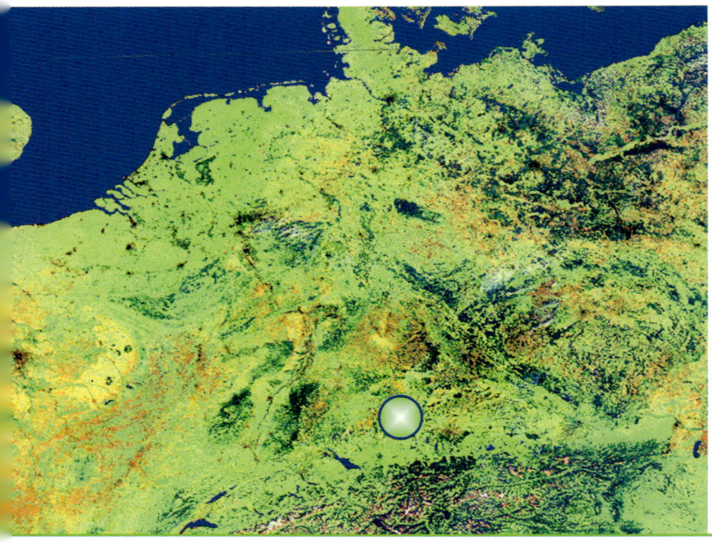

86 TOP IN THE BAVARIAN FOREST PARK THE GREAT PREDATORS, AMONG THEM BROWN BEARS, ARE HOUSED INSIDE VAST ENCLOSURES, WHICH ALLOW FOR EASIER VIEWING BY TOURISTS.

86-87 AND 87 THE BAVARIAN FOREST IS IN LARGE PART COMPOSED OF MIXED FORESTS, WITH VAST CONIFEROUS AREAS. THE PRIMARY FOREST HAS UNDERGONE MANY CHANGES OVER THE COURSE OF THE CENTURIES, AND MAINTENANCE OF THE WOODLAND AREAS IS TODAY THE OBJECT OF PARTICULAR CONCERN.

The Bavarian Forest National Park

Germany

The Bavarian Forest deserves the nickname "green inferno" not for its inaccessibility, but rather because in many ways it is similar to the primeval European forest. We are talking about a small fraction of the larger Bohemian Forest, which, in its entirety, extends over about 500,000 acres (200,000 hectares). The Forest, founded in 1970, was the first West German national park. Stretching north-south for 12 miles (20 km) and about 4 miles (6.4 km) across, the site conserves the primeval forest still standing in this western strip of Eurasia with all its extraordinary appeal and its mystery.

Among the different types of environments represented, besides the properly named "virgin" forest, there are also woods of large spruce and a forest of mixed broad-leaved and coniferous trees, as well as peat bogs. Closer examination reveals that the Bavarian Forest's environment is highly diversified; indeed, climbing toward higher elevations, in just a few miles one passes through landscapes that could also be encountered in traveling toward northern Europe.

A large part of the Bohemian Forest is mixed woodlands, in which the most typical phytosociological combination is that of silver firs, spruce, and beech trees. The presence of acidic soil, resulting from the breaking up of the rocky subsoil, fosters the growth of coniferous forests. At higher elevations, spruces are predominant, while at lower elevations, mixed woodlands with beech trees flourish. Overall, vegetation covers over 98 percent of the entire forest area. A total of 37,100 acres (15,000 hectares) – about half of the park's territory – have been left to themselves, with the result that bit by bit the Forest is transforming itself into something akin to "virgin forest," where man makes no impact and wilderness takes the upper hand. It is not uncommon to find large fallen trees in whose trunks grow mosses, lichens, and mushrooms. Inside these trunks live some of the most diverse species of life forms (above all invertebrates), recreating an ecosystem in miniature and contributing to the good health of the whole forest.

Noteworthy are what are truly considered to be the great "green sponges" – the peat bogs. Locally called *filze*, they are marked by an exceptional growth rate among the mosses that compose them. Thanks to the levels of rainfall in the zone, in some spots the tireless bog mosses have managed to overrun layers of plant life, one after the other, finally reaching the substantial thickness of 16 ft (4.8 m).

Animal life is equally abundant and diverse in the Bavarian Forest, with representatives from the main vertebrate categories that once inhabited the great European forest. Here as well, however, the absence of large predators has a marked effect on the maintenance of the biological equilibrium. In fact, the wolf and the lynx, the two biggest predators, have been extinct for some time. As a result, many species of hoofed creatures now pose serious problems because of increasing overpopulation. Even if they were successfully re-introduced, it is unlikely the wolf and the lynx would be able to repopulate the area of the Forest in a short time or redress the population balance. Ironically, these two missing predator species can be seen in the vast enclosures at the park entrance, where visitors can observe a representative selection of the typical fauna. The decision to create a sort of small zoo of local species was strongly criticized at the time, but this initiative has made it possible to limit the heavy inflow of tourists into the depths of the park itself and into its most delicate and sensitive areas. Unrestrained inflows would create obvious management problems and would also threaten the delicate ecological balance in this last remaining tract of European forest.

88 AND 89 BOTTOM HEAVY RAINS AND RELATIVELY CONTAINED TEMPERATURES ALLOW FOR THE GROWTH OF MOSSES AND BRYOPHYTES ON THE TRUNKS OF TREES AND THE FOREST FLOOR.

89 TOP THE EUROPEAN LYNX IS ONE OF THE MOST IMPORTANT PREDATORS IN EUROPEAN AND SIBERIAN FORESTS. HAVING BECOME EXTINCT IN THE BAVARIAN FOREST IN 1846, IT WAS REINTRODUCED DURING THE 1990s.

90 TOP SEVERAL SPECIES OF MAMMALS CAN BE ENCOUNTERED IN THE BAVARIAN FOREST. AMONG THEM DEER, ROE DEER, WILD BOAR, AND OTTERS. THE EUROPEAN WOLF, WHICH LIVES IN A FEW PROTECTED AREAS OF THE PARK, IS ONLY MINIMALLY PRESENT.

90-91 AND 91 THE BAVARIAN FOREST PARK, FOUNDED IN 1969, ENCOMPASSES A VAST TERRITORY COVERED BY DENSE WOODS. IN THE MOUNTAINOUS AREAS BETWEEN 3500 AND 5000 FEET HIGH (915 AND 525 M), THE CLIMATE IS HARSH AND THE PREDOMINANT TREE SPECIES IS THE SPRUCE.

92-93 DURING WINTER, SOFT LAYERS OF SNOW COVER EVERY INCH OF THE PARK.

The Bavarian Forest

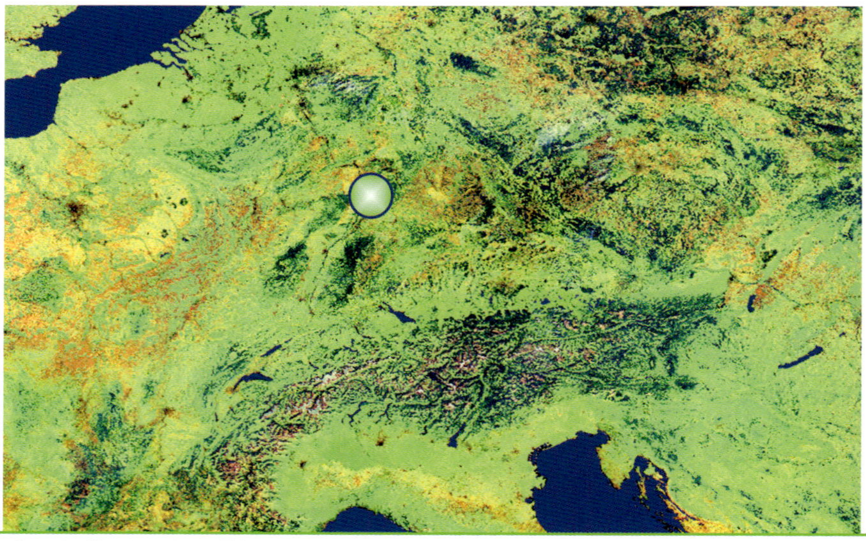

SAXON SWITZERLAND NATIONAL PARK - GERMANY - SA

Saxon Switzerland National Park

Germany

ON SWITZERLAND NATIONAL PARK - GERMANY - SAXON S

More than 200 years ago, some Swiss painters compared the characteristic landscape of the sandstone rock of the mountains along the Elbe River and south of Dresden to Swiss mountain scenery, and gave the German region the name "Saxon Switzerland." This region, located in southeastern Germany on the border of the Czech Republic, is justly famous for its particular Swiss-type landscape: in the Cretaceous period (135 to 55 million years ago), the oceans scraped romantic mountains out of the sandstone, creating bold peaks alternating with isolated and quiet gorges, making the area a true paradise for climbers. The Elbe River runs between gorges and canyons that offer shelter to numerous animal species, among them the otter, still relatively common here. As early as the 18th century, the landscape's romantic nature was a source of inspiration for painters and visitors. It is enough to recall that Caspar David Friedrich (1774-1840), one of the most famous landscape artists of all time, painted many of his pictures right in this area.

From the point of view of natural resources, Saxon Switzerland represents a unique example of a typical erosion landscape, formed during the Tertiary and Quaternary periods (between 65 million years ago and recent times) out of the sandstone dating back to the Cretaceous period. For this reason, the area was declared a national park in a protected transborder area. The park can claim, among other things, a highly diversified fauna and flora; in addition, it is an excellent area to go for bicycle rides or walks. Within the total area of 270 sq. miles (700 sq. km) are an exceptional number of sandstone boulders, mountain faces, spires, and a dense series of gorges and valleys can be seen.

Today the region is of great interest both to nature conservation specialists and to flora and fauna specialists. Since 1992, a census of the vascular plants on both the German and Czech sides of the park has been maintained, with the aim of developing a proposal for joint conservation measures and the production of a distributable atlas, together with a list of threatened species.

94 AND 95 SAXON SWITZERLAND IS CHARACTERIZED BY UNIQUE MORPHOLOGY: TALL SPIRES OF SANDSTONE, SMOOTHED BY A CRETACEOUS-ERA OCEAN, ALTERNATE WITH GORGES AND CANYONS, THROUGH WHICH RUNS THE ELBE RIVER.

96-97 THE VALLEY OF THE ELBE RIVER IS A LAND RICH IN HISTORY, CASTLES, PARKS, AND MYSTERIOUS WOODS. HOWEVER, ABOVE ALL THERE ARE ITS CANYONS, CROSSED IN SOME PLACES BY PICTORESQUE STONE BRIDGES, THAT GIVE IT AN ALMOST FAIRY-TALE-LIKE ATMOSPHERE.

Jasmund National Park

Germany

Rügen, the largest of the German islands, is located off the Mecklenburg-Western Pomerania region, in what was once the German Democratic Republic, or East Germany. Facing onto the Baltic Sea, it has an area of 361 square miles (935 sq. km) and is home, on its east coast, to the protected area of the Jasmund, the smallest but probably the most beautiful national park in Germany. Despite being only 10,000 acres (4050 hectares) in extent, it has both wetlands and dry meadows, not to mention its large woodlands. Its topography is characterized above all by large vertical chalk cliffs formed as a result of an upthrusting of the land during the Ice Age. Beyond Königsstuhl, which at 384 ft (106 m) is the highest point along this sheer, the cliff continues for a good 6 miles (9.6 km).

On Stubnitz, a high plateau formed during the Ice Age, the national park encompasses extensive beech woods dating back to the 13th century, but also many other plants that love wet conditions find their ideal environment here. Furthermore, the roots of alders have formed vast peat bogs around local fresh-water springs. This national park boasts among other things Germany's greatest variety of woodlands with their typical fauna and flora, full of numerous species. Among the former, there is the bee-eater, which nests in the chalk cliffs, and the golden eagle. Among the latter, there is bog moss and cotton grass, which is predominant in the peripheral swamps lacking trees, along with sundew and other plants that flourish in these rather than in calcareous and lime-rich soils.

98 BOTTOM AND 99 TOP RÜGEN ISLAND FEATURES A PARTICULARLY DIVERSE AND LUSH NATURAL SETTING, COMPOSED LARGELY OF EVERGREENS.

98-99 THE 360 MILES (480 KM) OR SO OF COASTLINE ON RÜGEN VARY GREATLY, WHERE WHITE, SANDY BEACHES ALTERNATE WITH DUNES, GREEN COPSES, SUDDEN CLIFFS, BAYS, AND LAKES.

100-101 THE WHITE CLIFFS OF RÜGEN HAVE INSPIRED POETS AND PAINTERS OVER THE CENTURIES.

Bialowieza National Park

Poland

The extensive Bialowieza Forest, straddling the border between Poland and Belarus, is the remnant of what was once the forest that cloaked all of Europe. The Bialowieza tract is actually one of the best preserved forest tracts on the Old Continent and, in many ways, represents to Europe what Yellowstone represents to North America, both in terms of wilderness and the animal and plant species to which it is home. Bialowieza extends over 346,000 acres (140,000 hectares) between Poland and Belarus, 143,000 (57,800 hectares) of which are on the Polish side. The plants that grow in the forest have no peers in the rest of Europe, both for their exceptional size and their age: majestic oaks alternate with bushes and dense forest-floor vegetation, with clearings that have formed as a result of the fall of centuries-old trees. The forest is home to many particularly big trees, estimated to be hundreds of years old, and to many different plant and animal species, some of them – the European bison, the wolf, and the lynx – quite rare. In Bialowieza, 900 species of vascular plants (of which 26 are tall-trunk trees), 12 species of orchids, 254 lichens, 80 bryophytes (mosses and worts), and 1500 mushrooms have been catalogued. The Bialowieza Forest preservation measures have quite distant origins: as early as the 19th century, the Polish kings had limited the hunting of large animals. Thereafter, following independence in 1918, Poland instituted important conservation measures. Three years later, the first forest reserve was established – in Bialowieza – bringing over 11,000 acres (4450 hectares) of forest under partial protective regulations. In 1932, Bialowieza was declared a national park. In many ways, the Bialowieza Forest reflects the larger Polish landscape inasmuch as 28 percent of the nation's land is covered by woods. In 1977, Bialowieza Forest became a Man and Biosphere Reserve, and in 1979, it was inscribed in the World Heritage List, the list of sites around the world that should remain unchanged inasmuch as they are the heritage of future generations. Seven zones are today recognized as biosphere reserves, eight marshes are protected by the Ramsar Convention for swamp and marshlands (an international conservation agreement

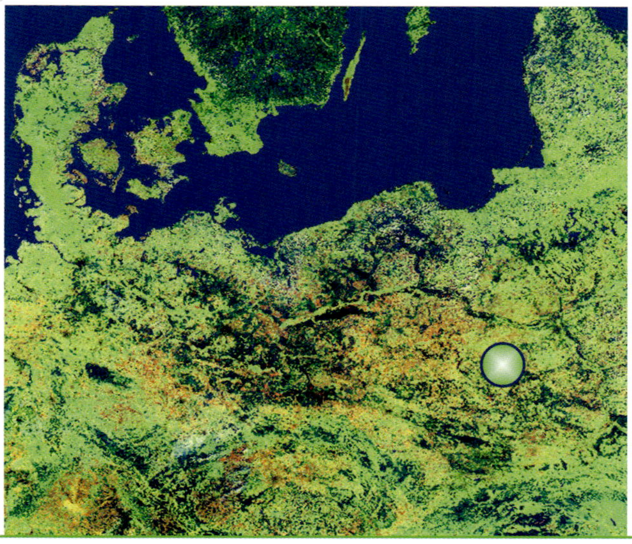

TIONAL PARK - POLAND - BIALOWIEZA NATIONAL PARK -

signed at Ramsar, in Iran, in 1971), and four other sites have received the title of biosphere reserves.

The Polish part of the Bialowieza Forest is home to over 60 species of mammals, including large carnivores like the wolf and the lynx and herbivores, the most famous of which is the European bison. The now stable bison population is the result of one of the rare successful projects of breeding a species in captivity: Bialowezia's last wild bison was killed by poachers in 1919. In fact, in the 1920s, the species was close to extinction; only about 50 bison remained, distributed among zoos around the world. Through the careful selection of 13 individuals, it was possible to begin the reintroduction of the species. In Bialowieza, this reintroduction took place in 1929, in one of a few monitored enclosures, and then, in 1952, out in the wild. In Poland, Lithuania, Ukraine, and Russia reintroduction occurred during the 1950s. Overall, the process has been particularly positive in its various sites and the results are very encouraging. Today Europe has a bison population of about 3200 specimens, of which some 400 graze in Bialowieza Forest.

At the other end of the size scale, the forest also has an impressive insect population estimated to number 10,000 species, including some rare ones dependent on very old trees.

102-103 BIALOWIEZA IS ONE OF THE LAST VESTIGES OF THE DECIDUOUS FOREST THAT ONCE WAS WIDESPREAD IN MUCH OF EUROPE.

102 BOTTOM BIALOWIEZA IS ONE OF THE FEW PROTECTED AREAS IN WHICH EUROPEAN BISON HAVE SURVIVED.

103 TOP THE PARK'S GREAT SIZE AND WILD FORESTS MAKE IT A HAVEN FOR SEVERAL LARGE MAMMALIAN SPECIES, INCLUDING THE BISON AND THE EUROPEAN LYNX.

104-105 AND 105 IN BIALOWIEZA, SOME AREAS FLOOD PERIODICALLY, PROMOTING INCREASED ANIMAL LIFE AND A REMARKABLE VARIETY OF VEGETATION.

106-107 BIALOWIEZA'S LARGE DEER POPULATION IS AN IMPORTANT WILD LIFE ACHIEVEMENT; THE DEER ALSO ARE A FOOD SOURCE FOR WOLVES AND LYNXES.

Bialowieza

The Cliffs of Étretat

France

According to some, it is at the cliffs of Étretat, in Normandy, that temperate Europe ends and the "Great North," the harbinger of leaden skies, cold seas, and damp, constant, and persistent winds, begins. In fact, the wind is one of the most important characteristics of this area, a humid wind that blows tenaciously, shaping the rocks and thereby sculpting the cliffs.

The landscape of Étretat is truly breathtaking: the cliffs, over 300 ft (90 m) high in places, have little to envy in those of the Mediterranean. Even though they are not as dazzling as those well-known cliffs of Bonifacio in Corsica, no others are covered by such a cloak of vegetation as are these. Giving its name to this coastal tract is Étretat, a seaside village of 1600 inhabitants situated on the "Alabaster Coast," which extends from Le Havre to the mouth of the Bresle River, in other words along the entire coastline of the department of the Seine-Maritime. Along the 87 miles (140 km) of its shoreline, the Alabaster Coast opens onto the sea with coastal dales and wide valleys that are of two different types: the wet ones, crossed by rivers with a sizeable flow, and the dry ones, of a karstic nature, which hide underground waterways. Not far from the southernmost end of this coastal tract, the beach of Étretat, well known for its characteristic pebbles, forms an ample natural amphitheater bordered by spectacular "gates," natural arches sculpted by erosion that rise suddenly to the left and right of the beach. The left side is bordered by the so-called Aval Gate, which recalls the slender ogival form of a pointed arch, typical of French cathedrals. From the top of the crest of this cliff, a rocky arch that is brusquely interrupted, the view is gorgeous: from there, it is possible to admire the massif arch of Manneporte, characterized by a monumental natural structure, and also the Aiguille made famous by Maurice Leblanc, the creator of the gentleman-thief Arsène Lupin. Leblanc made the Aiguille, a large, natural obelisk that stands isolated and is about 230 ft (70 m) high, his character's secret hideout. So amazing is the beauty of these cliffs that they have stimulated the imagination of great personalities from cultures of every period, like the painter Claude Monet and the immortal Victor Hugo, according to whom the cliffs of Étretat are the "greatest architectural work" in existence. On the other end of the beach rises the Amont Cliff, on whose crest stands a chapel dedicated to Notre-Dame-de-la-Garde. The Aiguille and the Aval Gate are the result of active

108-109 AND 109 TOP UNMISTAKABLE SYMBOLS OF THE CLIFFS OF ÉTRETAT ARE ITS CHARMING NATURAL WINDOW, THE SOLITARY SPIRE OF THE AIGUILLE, AND THE MANY SEA BIRDS THAT NEST THERE.

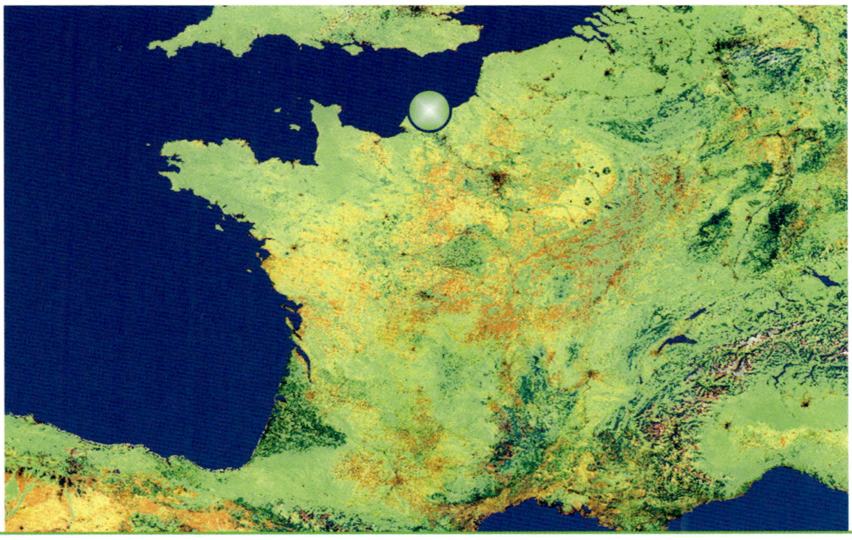

wind erosion on the cliffs. The calcareous rock faces are, in fact, very friable and they literally retreat little bit by little bit with every stormy assault of the sea, which vigorously shapes them. The formation of these natural "monuments" is explained by the fact that their stone composition is more solid and more resistant than that of other cliffs. The sea and the wind regularly pit themselves against "prey" that, in the end, is "easy." Launching pebbles and rocks against the chalk walls, the sea carves and punctures them, creating caves and gorges. The Alabaster Coast boasts pebble beaches, with the pebbles extruded from their chalk matrix by the erosive force of the wind. As the wind blows, these pebbles, bit by bit, are loosened and extracted from the cliffs. Thereafter, the sea gives them their characteristic smooth shape. The waters that infiltrate between the cracks then complete the erosive process, forming spires or arches full of charm and mystery.

110-111 AROUND CAUX THE CHALK CLIFFS RUN FOR SEVERAL MILES AND ARE OFTEN UP TO 300 FT (90 M) HIGH. BY 1900, ÉTRETAT HAD BECOME A FASHIONABLE RESORT; TODAY IT CONTINUES TO DRAW MANY VISITORS.

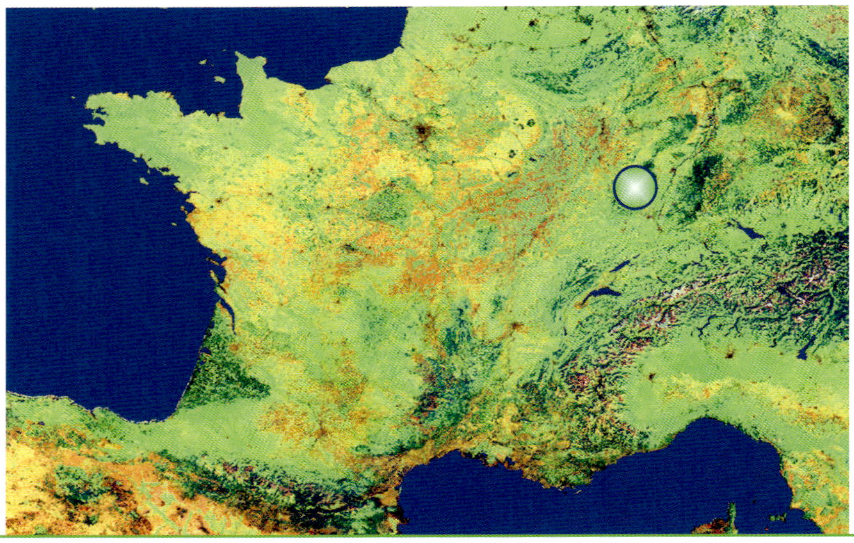

THE REGION OF THE VOSGES MOUNTAINS - FRANCE - TH

112 TOP THE MALE HAWFINCH HAS BROWN FEATHERS WITH REDDISH HIGHLIGHTS, RED HAUNCHES, AND DARK WINGS. IT CAN BE DISTINGUISHED FROM THE FEMALE BY ITS STRONGER COLORING AND ITS DARK BROWN, ALMOST RED, HUED HEAD.

112-113 AND 113 THE VOSGES REGION IS FULL OF LAKES. MANY AREAS ARE DISTINGUISHED BY THE PRESENCE OF SWAMPS AND VAST SEDGE MEADOWS, WHICH DEVELOP IN AREAS MARKED BY PEAT BOGS.

The Region of the Vosges Mountains France

The Vosges chain, located in central Western Europe, extends along the western side of the Rhine valley in a north-northeast direction, from Basel, Switzerland to Mainz, Germany. It is interesting to note that there is a remarkable resemblance between the environment of the Vosges and that of the Black Forest in Germany, extending along the other bank of the Rhine. Both areas share the same conformation and are characterized by the presence of abundant forests, dominated, at lower levels, by pasturelands and rounded peaks of much the same height. To speak about the Vosges also means speaking about the Rhine: the river's "corridor," a vast valley running north-south, was formed following a surface-level fragmentation of the earth's crust, which took place about 40 million years ago due to the spreading of the continental plates. Over the course of millions of years the river has deposited and continues to deposit sediment along its way toward the North Sea, creating an alluvial plain between the Vosges and the Black Forest, with the forests and fields regularly fertilized and irrigated by floods. The region contains two protected areas: the Regional Natural Park of the Northern Vosges and the Ballon des Vosges Regional Natural Park, both dedicated to the conservation of a natural setting characterized largely by a succession of vast woodlands, peat bogs, and tall cliffs. The Northern Vosges park was founded in 1975, and by 1989 – together with the German Naturpark Pfälzerwald, which protects the zones bordering on the German Rheinland-Pfälz region – is an important cross-border biosphere reserve – as UNESCO now classifies the two-part area.

In total, within the entire massif, about 400 peat bogs have been counted; they are a highlight and strong point of the Vosges environment. The development of these very special environments, which grew out of the granite basins left by the retreating glaciers, has been fostered by the cool and very rainy climate. Above 3000 feet (915 m), the pine tree, more sensitive to wind and freezing temperatures, gives way to the fir tree.

The fauna of the Vosges is typical of central Europe, with many birds, above all birds of prey, and a certain number of small- and medium-sized mammals. In this region too, wolves were brought to extinction in the past, but recently a reintroduction program has been set up in the Northern Vosges area. Another great predator, the lynx, had also been brought to the verge of extinction and has recently been re-introduced. The project was conceived in 1972 and began in 1983, with the release of three lynxes from the Carpathians.

Vosges

114-115 IN WINTER, THE SWAMPS FREEZE, WHILE THE MOUNTAINS APPEAR IN EXTREMELY CHARMING SHADES OF LIGHT AND SHADOW.

115 THE VOSGES MOUNTAINS ARE TRULY AN AREA OF GREAT IMPORTANCE TO NATURE LOVERS. EVERY WINTER, THE LANDSCAPE IS CLOAKED IN SNOW, AND BECOME TRULY APPEALING TO THE VISITORS.

116-117 IN FALL, THE LANDSCAPES OF THE VOSGES ARE OFTEN MARKED BY MISTY, HAZY HUES THAT GIVE THE COUNTRYSIDE A MYSTERIOUS FAIRYTALE APPEARANCE.

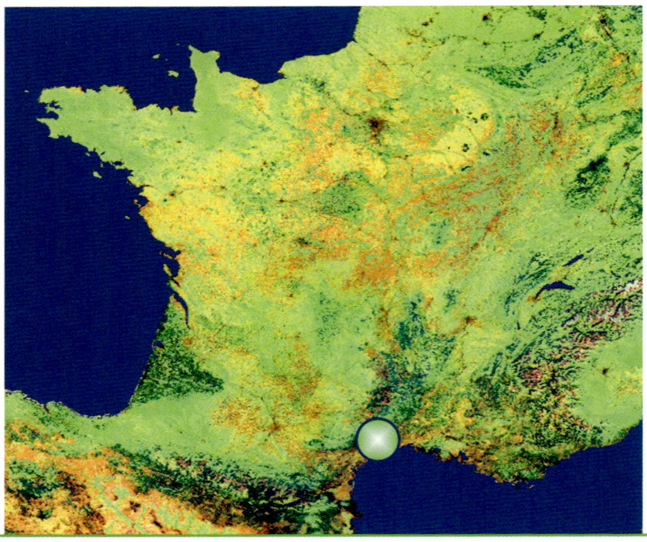

118 TOP THE STRONG WINDS THAT BLOW ALONG THE COASTS OF SOUTHERN FRANCE CREATE SHIFTING DUNES.

118-119 AND 119 IN THE CAMARGUE, THE COLORS, THE FAUNA, AND THE FLORA ARE UNIQUE. IT IS A TRUE UNION BETWEEN THE ACTIVITIES OF MAN AND NATURE IN THE WILD.

120-121 ONE OF THE CAMARGUE'S GREATEST ATTRACTIONS IS THE EQUINE BREED NAMED AFTER THE AREA. IN FACT, IN 1928 THE RESERVE WAS CREATED TO PROTECT THESE HORSES.

THE REGIONAL NATURE RESERVE OF THE CAMARGUE - F

The Regional Nature Reserve of the Camargue

France

The Camargue is a truly surprising landscape, covered with boundless carpets of glasswort ranging from green to ochre depending on the seasons, and swept by constant winds (the well-known *mistral* from the north and the *maestro* from the northwest).

The region is a true paradise above all for birds that live in limicolous (muddy) environments and for the many animal species that live in water, from small invertebrates to fish and amphibians. Forged by the force of the sea and the action of the wind, the Camargue is a typical river delta, with a wild and changeable look. It includes marshes and valleys, not to mention vast beaches shaped by the tides. In all, the region is an ecological combination unique in France – and one that has no equals in Europe except possibly for the deltas of the Danube, the Guadalquivir, or the Po.

The Camargue, which was designated a park in 1975, encompasses an area of about 32,412 acres (13,116 hectares) around the outlet of the Rhône River, delimited by the mouths of the river (the Great Rhône and the Little Rhône) respectively located to the east and west, by the city of Arles to the north and by the shoreline of the Mediterranean to the south. The great diversity of the Camargue's habitats is above all to be found in the Rhône's long history of sedimentation, which bit by bit conquered the sea, depositing sand, creating lagoons, stagnant pools, and a multitude of fossilized sandbars.

The biological element enters this changeable background with ease: thanks to the constant action of the wind, vast areas of sand and coastal dunes formed, mixed with areas of vegetation able to survive and often prosper in a salty environment of glasswort on expanses of *garigue* (typical areas of Mediterranean scrub, with shrubs alternating with bare areas). All around stretch open meadows and steppe-like terrain, not to mention remnants of the original Mediterranean-scrub-type vegetation. The fauna is the kind typical of salty Mediterranean zones, but the Camargue is best known for the fowl (or ornithofauna) living there: over 270 bird species, 26 of which are of international importance, and almost the entirety of the Mediterranean flamingo population, with 8000 pairs. Abundant in plant and invertebrate species and protected from the interference of man, this habitat offers birds (which run the place!) a fantastic spot to rest and a safe refuge where they can nest.

The Camargue

122 BIRDS LIFE IS UNDOUBTEDLY IS ONE OF THE CAMARGUE'S MAJOR ATTRACTIONS, AND FLAMINGOES ARE ONE OF ITS MOST FASCINATING SPECIES. THE RESERVE IS A MAJOR NESTING SITE FOR THIS HANDSOME BIRD.

123 THE FLAMINGOES' PINK HUE DERIVES IN PART FROM THEIR DIET OF CRUSTACEANS AND ALGAE RICH IN CAROTENE.

124 TOP THE ABUNDANCE OF SMALL- AND MEDIUM-SIZED FAUNA HAS FAVORED THE PRESENCE OF PREDATORS LIKE THE RED FOX, ABUNDANT IN ALL AREAS OF THE CAMARGUE.

124-125 EXTENDING OVER ABOUT 27,000 ACRES (110,000 HECTARES), THE SALT MARSHES OF THE GIRAUD, ADJOINING THE GREAT RHONE ESTUARY, ARE THE MOST IMPORTANT IN THE CAMARGUE. THE GIRAUD AREA'S SALT PRODUCTION IS ABOUT 800,000 TONS (812,840 TONNES) PER YEAR.

125 TOP THE DUNES OF THE CAMARGUE ARE MARKED BY VAST GLASSWORT MEADOWS. THIS PLANT IS TYPICALLY HALOPHYTIC, MEANING ADAPTED TO BRACKISH CONDITIONS.

125 BOTTOM THE CAMARGUE IS ONE HUGE SWAMP, WITH ABOUT 300 SQUARE MILES (780 SQ. KM) OF RIVER AND BRACKISH-WATER ECOSYSTEM. THE REGION IS A MYRIAD OF FRESHWATER AND SALTWATER PONDS, HOME TO NUMEROUS SPECIES OF MARSH BIRDS.

126 AND 127 BETWEEN MARSEILLE AND CASSIS IS THE FAMOUS CALANQUE OF EN-VAU, A 2-MILE (3.2-KM) DEEP CALCAREOUS MASSIF EXTENDING FOR OVER 12 MILES (20 KM) ALONG THE SEA FRONT. THE MASSIF ATTRACTS MANY FREE CLIMBERS.

128-129 THE SHORES OF THE CALANQUES HAVE A DOZEN OR SO UNDERWATER CAVES AND SEVERAL LITTLE INLETS PROTECTED FROM THE FIERCE WINDS AND TIDES.

The Calanques of the Côte d'Azure

France

In the south of France, within an area of about 13,600 acres (5500 hectares), is the one-of-a-kind attraction of the white-walled *calanques*; impressive fjord-like inlets that cut through the cliffs that characterize the entire coast between Cassis and Marseille (the Marseilleveyre massif) and face the Frioul and the De Riou islands. Along the coast, magnificent white cliffs drop sheer into the sea, marked by deep coves.

The most famous are the Calanque de Sormiou and the Calanque d'En-Vau, both of which project power and beauty. The *calanque* of d'En-Vau, about 490 ft (150 m) wide, retreats about 2000 ft (610 m) into the land, enclosed between two sheer walls of rough rock and narrowing until it ends in a small beach of white sand, forming a veritable southern fjord. The appearance of this landscape and seascape was a powerful factor in the evaluation of the importance of the *calanque* region, but a significant environmental element, one which makes the area a true environmental gem of the European Mediterranean, must not be overlooked. A study of the *calanque* habitats done under the auspices of Europe's Natura 2000 project resulted in the cataloguing of 26 natural environments, to which can be added 40 mixed habitats. Within the *calanque* region about 900 plant species have so far been identified, including 660 species of vascular plants and 83 species that are protected on a regional, national, or international level. In terms of fauna, the *calanque* region contains communities well adapted to the extreme habitat and dryness of southern France, with some outstanding examples such as Bonelli's eagle (*Hieraaetus fasciatus*) and several species of seabirds. Some 30 percent of the population of Cory's shearwaters (*Calonectris diomedea*) and storm petrels (*Hydrobates pelagicus*) in France are found on this massif, as are 10 percent of the Mediterranean, or Yelkouan, shearwaters (*Puffinus yelkouan*). Local species of reptiles number 12, with 4 amphibian reptiles; among mammals, chiropters (bats), which find shelter in the region's numerous karstic caves, hold particular importance. Regarding marine biodiversity, the sea is home to large meadows of posidonia, or Mediterranean tapeweed, and is one of the high-priority habitats included in Natura 2000's Habitat Directive as an ecosystem of great value to the Mediterranean coast.

One natural feature in particular merits mention, namely Cosquer Cave. Discovered only a few years ago, the cave is renowned for the evidence it contains of prehistoric man: the rock paintings it holds are, in fact, comparable in interest and value to those of Lascaux or the Combe d'Arc. The cave contains two series of drawings, the first of which dates back to 20,000 to 17,000 years ago, and the other to 29,000 to 26,000 years ago, periods that correspond to the two most recent phases of the last Ice Age. Today, the entire *calanque* area and the De Riou Islands, which received their *site classé* registrations in 1975, represent a "green lung" for the Marseille metropolitan area, a fact that has helped promote their expanded recreational use. Given that an estimated one million people a year visit the *calanques* and the islands, it is easy to appreciate the problems that can occur in areas in which restrictions are already in place to ensure successful conservation. Fortunately, the *calanques* are among the "sites of communal importance" that Natura 2000 has listed within the framework of the Habitat Directive. In addition, it is recognized that the conservation measures in effect in the area are clearly insufficient. In fact, many environmental associations are now demanding that the *calanques* area be designated a national park, an action that would. It is hoped, have the effect of ensuring greater and more efficient protection and safeguarding of the natural environments and their biodiversity.

The Nature Reserve of Bonifacio

France

THE NATURE RESERVE OF BONIFACIO - FRANCE - THE NA

Clinging to its white cliffs, the city of Bonifacio, in southern Corsica, gives its name to one of the Mediterranean's most evocative places, the so-called "Mouth of Bonifacio," which is a broad, deep bay. There, Corsica, in antiquity known as Kallisté (meaning beautiful), faces nearby Sardinia, projecting out over a swathe of crystal-blue sea dotted with islands and islets, the natural geographical continuation of which is found in the Maddalena Archipelago, which lies just off Sardinia. Many of what are truly the most fascinating parts of the Mediterranean are in this area.

In terms of physical and geographical factors, the flora and fauna of these Mediterranean outcrops represent a single entity even though in political terms the environment and its ecosystems are "shared" by Corsica and Sardinia, islands that are respectively part of France and Italy. For this reason, an international (or cross-border) protected park has been recently been created that includes among its goals the safeguarding and environmental management of the strait separating these two large islands.

The resulting entity, the Park of the Mouth of Bonifacio, can rightfully claim an exceptional environmental and naturalistic heritage encompassing several archipelagos and many islands. The park is home to unique coastal and marine ecosystems, which in esthetic terms are the equivalent to incomparable landscapes. Among these is the Lavezzi Archipelago and Cerbicali Islands, the Bruzzi (or Monks) peninsula, the cliffs of Bonifacio, and the tidal pools of Ventilegne, not to mention the Three Swamps of Suartone (an ecological and floral site of both Mediterranean and European importance), the beaches, the tidal pools, and the marshes of Rindinara, the terrain of the Conservatoire du Littoral, and the National Park of the Maddalena Archipelago. The marine areas (which are islands and their waters) that the international park now protects number six. Three are on the Corsican side (the islands of Lavezzi, Scandola, and Finocchiarola) and the other three are on the Sardinian side (the islands of La Maddalena, Asinara, and Tavolata). The Lavezzi Archipelago, extending over 10,900 acres (4411 hectares), is bounded by the Capicciolu and dello Sperone headlands and by the Sperduto Islands, off the Corsican coast.

At an earlier time, Bonfacio was considered only as a place of botanical riches and abundant birdlife, but today the port town is recognized as an extremely important center both for fish species and for marine biodiversity, characterized by the presence of all the Mediterranean species of dolphins, whales, sea turtles, and many other important species. Giant limpets, a species all but extinct in the Mediterranean because of coastal exploitation and pollution, are found there. A great variety of fish, soft corals, sea fans, as well as vast meadows of Mediterrean tapeweed are also present. Naturally, marine bird species are numerous; they include cormorants, Atlantic puffins (*Fratercula arctica*), and Audouin's gulls (*Laurus audouinii*). Reptiles too boast diversity, with three species of geckos (*Tarentola mauritanica, Hemidactylus turcicus,* and *Euleptes europaea*). The founding of the international park was a great challenge for the conservation movement, and meant overcoming national policies and state border concerns and even international issues, all in the interests of guarding wildlife. The park, in addition to actively protecting an area distinguished by significant biological diversity and a variety of landscapes, has to attempt pragmatically to regulate the access of international oil corporations to the region, not to mention indiscriminate tourism and an aggressive and intensely active fishing industry.

130 AND 130-131 HIGH CLIFFS EXTEND ALONG SOME 15 MILES (24 KM) OF THE COAST.

131 TOP THE UNCHANGING BEAUTY OF BONIFACIO, NICKNAMED "BONIFAZIU LA BLANCHE," STRIKES EVERY VISITOR.

132-133 BONIFACIO'S CLIFFS ARE REMARKABLE FOR THEIR GRANDEUR. THEY ALSO ARE A TRUE TRIUMPH FOR THE CONSERVATION OF MARINE BIODIVERSITY; IN TERMS OF RARE SPECIES THE LOCAL WATERS ARE UNRIVALED THROUGHOUT THE MEDITERRANEAN.

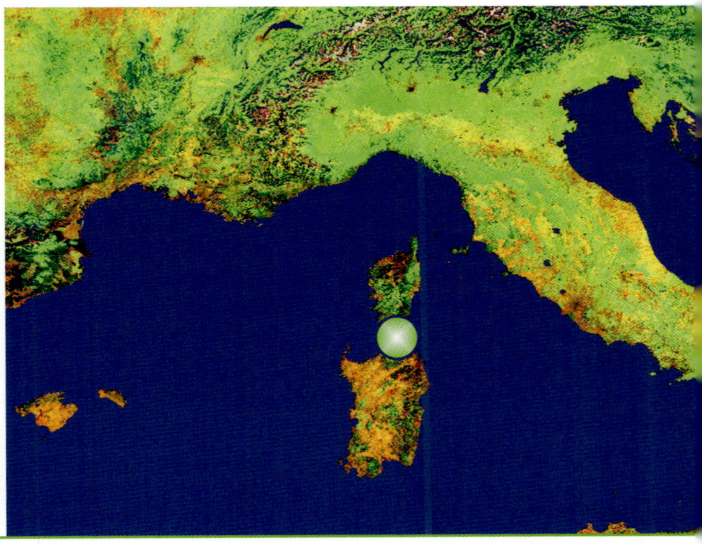

134 TOP THE GRIFFON VULTURE IS THE SYMBOL OF MONFRAGÜE.

134-135 NUMEROUS DAMS MAKE THE TAGO RIVER LOOK LIKE AN ARTIFICIAL LAKE.

135 TOP THE ROCKY STACK OF SALTO DEL GITANO IS AN IDEAL NESTING SITE FOR MANY BIRDS OF PREY.

135 BOTTOM THE ARID CLIMATE AND THIN SOIL HAVE GIVEN RISE TO THE DEHESA TYPE OF VEGETATION, AN ECOSYSTEM DIFFERENT FROM THAT OF CULTIVATED AND WILD AREAS.

Monfragüe National Park

Spain

The flight of the cinereous vulture (so-named because of its ash-colored head, which has distinctive black patches), griffon vulture, and many other large birds of prey is nature's signature above Monfragüe National Park, a well-known and much-loved Mediterranean area, crucial to European conservation efforts. Monfragüe encompasses a vast portion of environmentally well-preserved Mediterranean mountains, which helps to make it possible for a notable variety of animals and plants to survive. The combination of large open spaces and areas covered in dense, almost impenetrable woods, not to mention pasturelands, promotes this diversification. Nonetheless, Monfragüe is above all identified by the famous and impressive crag of Peña Falcon, which is home to one of Europe's most populous colonies of griffon vultures. On this rocky wall, however, many other rare or difficult-to-observe birds nest, like the black stork, the Egyptian vulture, and the peregrine. From Monfragüe Castle it is possible to catch an occasional glimpse of these birds of prey in flight. The observation post overlooks the whole of Peña Falcon, and the flights of the griffon vultures, accompanied every now and then by an Egyptian vulture, always delight the watcher. However, in Monfragüe, as in the majority of European parks in which great birds of prey are found, man's help is indispensable in maintaining the populations. In the castle's vicinity are also black wheatears, rock buntings, azure-winged magpies, spectacled warblers, Sardinian warblers, woodlarks, and rock sparrows. At selected points in the park, man-made observation posts have been set up that provide excellent opportunities for nature-watching. It is also possible to observe the common stork nesting in trees, whereas in the nearby town of Trujillo (as in many other European towns), the storks nest on the house-roofs. However, not only griffon vultures and other birds can boast of living in Monfragüe. In the wetter and shadier zones, important tracts of primeval Mediterranean woods survive. Among the more prized plants are the cork tree and majestic speciments of the deciduous oak *(Quercus faginea)*, the Montpelier maple *(Acer monspessulanum)*. The forest floor is composed of a mix of *Lauraceae* (for example, the laurel and the terebinth), plants that were witness to ancient times when this type of forest was predominant in Europe and in the borderlands of the Mediterranean basin.

Among the mammals in the Monfragüe, the Iberian lynx deserves mention: the park is one of the last bastions of this animal. Among the great variety of amphibians, the Mediterranean tree frog and the Boscà lizard should be remembered. Still very abundant are the Mediterranean pond turtle (*Mauremys leprosa*) and the European pond turtle (*Emys orbicularis*), found here wherever there is a pond or lake. Finally, it is worth remembering how indispensable to the park's equilibrium is *dehesa*, a farming model still widespread in Spain and northwestern Africa. The *dehesa* model is common on vast tree-covered savannas, creating an alternating pattern of open spaces normally used for grazing herds of livestock and trees (above all cork and holm oaks), from which acorns, wood, cork, and charcoal are obtained.

136 TOP AND 136-137 MONFRAGÜE IS ONE OF EUROPE'S MOST CHARACTERISTIC AND BEST PRESERVED ZONES OF MEDITERRANEAN SCRUB AND SCLEROPHYLL-TYPE FOREST.

137 TOP TYPICAL SPECIES LIKE THE RED FOX LIVE IN MONFRAGÜE.

137 BOTTOM BESIDES BIRDS OF PREY, TWO DIFFERENT SPECIES OF SWAN, THE WHITE SWAN AND THE BLACK SWAN LIVE IN THE PARK.

Monfragüe

138 TOP LEFT IN SPRING AND FALL, DEER THAT NORMALLY FREQUENT THE EVERGREEN FORESTS, WILL MOVE TO LOWER LEVELS, OFTEN CLOSE TO POPULATED AREAS. IN SUMMER THEY MAY MOVE UP ABOVE THE TREE LINE.

138 TOP RIGHT AND 138-139 LAKES, STREAMS, WATERFALLS, AND EVERGREEN WOODS ARE AN IDEAL ENVIRONMENT FOR MANY BOTHANICAL SPECIES AND FOR ANIMALS OF ALPINE OR NORTHERN ORIGIN.

The Pyrenees

Spain

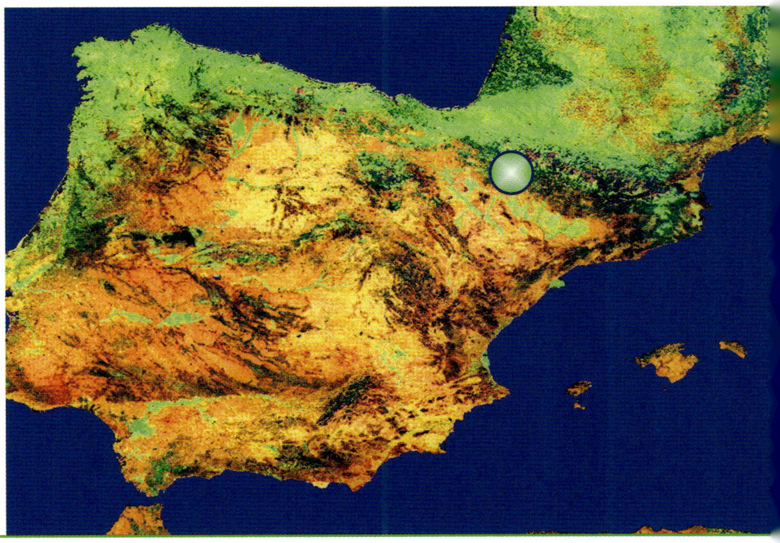

The long chain of the Pyrenees, a true natural geographical border between France and Spain, represents a shrine to naturalistic, cultural, and landscape treasures of great importance to the entire European continent. From a geomorphological point of view, the Pyrenees are older than the Alps, to which they are often compared, despite the latter being relatively "recent." The peculiar nature of the Pyrenean chain is due to the abundance of granite, a rock particularly resistant to erosion, and also to the occurrence of less glacial activity than in the Alps, which has helped to preserve the Pyrenees's rough outlines. The eastern portion of the chain is largely composed of granite and gneiss, whereas in the western part, the granite peaks alternate with layers of limestone, a component that guarantees the presence of an elaborate karstic system that is particularly extensive and interesting.

The east-west direction of the chain favors great differences and uniqueness from a climatic point of view as well. The northern slopes actually receive a larger amount of rain, while the southern ones have a more steppe-like appearance. More particularly, however, the Pyrenees can be divided along their length into three climatic sectors. The western tract features a humid oceanic climate, the central section is characterized by a continental climate (hot summers and tough winters), while the eastern sector is subject to the Mediterranean influence, with hot summers and rainy autumns. The glaciers are confined to the northern slopes of the central Pyrenees and, unlike Alpine glaciers, do not descend much into the valleys but rather feature their greatest expansion across the mountain summits. The consequences of these climatic differences are also reflected in the range's phytosociological state; i.e., the distribution of plants in the three major regions and their organization into vegetation "communities." Typically, along the western slopes the flora recalls that of central Europe, whereas in the eastern zone it is more distinctly Mediterranean. Conservation efforts are extensive on both sides of the Pyrenees. On the French side is the Western Pyrenees National Park; on the Spanish side are four national parks: Pyrenees, Alto Pallars and Aran Valley, Ordesa and Mount Perdido, and Cavadonga. Ordesa Park is located in the Aragonian Pyrenees in Huesca province, at the base of the massif of the Three Sisters, which, along with Mount Perdido, are the tallest peaks in the Pyrenees. The area is characterized by the presence of canyons and recalls the landscapes of the American Wild West. The two contiguous parks, Ordesa on the Spanish side and Western Pyrenees National Park on the French side, are an excellent example of a cross-border protected area. Just like many other mountain chains, the Pyrenees also boast many endemic *taxa* (systematic categories into which the various living species are classified) of animals and plants. The "jewel" in this case is the Pyrenean chamois (*Rupicapra pyrenaica pyrenaica)*, recently at the edge of extinction but today existing in generally abundant numbers. Unfortunately, the Pyrenean ibex (*Capra pyrenaica*) met a less favorable fate; the last specimen died in the early days of 2000. The marmot, extinct in the region since the end of the last Ice Age, has now been successfully reintroduced to many areas in the Pyrenees. In the same way, the brown bear, another mammal pushed to the edge of extinction at the beginning of the 1990s, was reintroduced in 1996. Among small-sized mammals, it is worth mentioning the Pyrenean desman (*Galemys pyrenaicus*), an elusive and highly rare insect-eating aquatic animal, present only in a few streams on the northernmost slopes.

The Pyrenean region is covered by tall trees; among them the beech and the silver fir stand out. In many areas, there are moors of noteworthy size; on the alpine horizon, pioneer vegetation (hardy, self-seeding plants) grows among the rock fields and meadows, whereas in the sub-alpine region, Swiss mountain pines are prevalent. Beech and fir-tree forests extend between 3000 and 5300 feet (915 and 1615 m); farther south, at the same altitude, are forests of *Pinus sylvestris*. Among the endemic plant species is the peculiar Pyrenean snapdragon (*Ramondia pyrenaica*), a plant belonging to the Gesneriaceae, a family generally found in tropical zones, and present in Europe only in this form.

140 TOP THE PYRENEAN CHAMOIS IS OVERALL MORE AGILE THAN THE ALPINE CHAMOIS, AND HAS AN ENVIABLE GRACE.

140-141 THE HIGH LEVEL OF RAINFALL MAKES THE PYRENEES A CHAIN WITH PARTICULARLY ABUNDANT VEGETATION. A CLEAR DIFFERENCE IN CLIMATE CAN BE NOTICED BETWEEN THE EASTERN PYRENEES, WITH THEIR DRY ENVIRONMENT, EXTREME TEMPERATURE VARIATIONS, AND AUTUMN RAINS, AND THE REST OF THE CHAIN, WHICH FEATURES A MARITIME CLIMATE, WITH HIGH HUMIDITY AND LIMITED TEMPERATURE VARIATION.

141 TOP THE PROFILE OF THE PYRENEES MOUNTAINS ARE PARTICULARLY HARSH, DUE TO THE HEAVY PREVALENCE OF GRANITIC ROCK AND TO MINIMAL GLACIAL ACTIVITY, WHICH NEVER SMOOTHED THEIR ORIGINAL SEVERITY.

141 BOTTOM UNLIKE IN THE ALPS, WHICH ARE NOTED FOR LARGE LAKES, THE PYRENEES HAVE MUCH SMALLER LAKES.

142-143 THE PYRENEES ARE HOME TO SEVERAL PROTECTED AREAS, OF WHICH THE MOST FAMOUS IS THE NATIONAL PARK OF ORDESA AND MONTE PERDIDO. HERE, UNTIL THE END OF THE 20TH CENTURY, THE NOW-EXTINCT PYRENEAN IBEX SURVIVED.

The Pyrenees

144 THOUGH THE LARGER MAMMALS HAVE BY NOW LARGELY DISAPPEARED FROM THE PYRENEES, SMALLER ONES, LIKE FOXES, ARE STILL NUMEROUS.

144-145 THE FRENCH SIDE OF THE PYRENEES IS WELL WATERED, AND ENDEMIC AQUATIC SPECIES INCLUDE THE PECULIAR DESMAN AND THE PYRENEAN BROOK SALAMANDER.

145 TOP THE ORBAIETA WOODS, AT THE 2500-F (760 M) LEVEL, EXTEND ALONG THE BANKS OF THE IRATI TORRENT. THE ORBAIETA TRACT IS ONE OF EUROPE'S BEST-PRESERVED FIR FORESTS, WITH VAST PRISTINE AREAS SIMILAR TO PRIMEVAL FOREST ECOSYSTEMS.

146 TOP PRAIA DA ROCHA IS ONE OF THE BEST-KNOWN AREAS IN THE ALGARVE.

146-147 THE NATURE RESERVE OF THE SUDOESTE ALENTEJANO AND THE VICENTINA COAST IS 184,809 ACRES (74,790 HECTARES) IN EXTENT.

147 TOP ALONG THE COASTLINE OF VILA DO BISPO, NATURE HAS COMBINED WITH THOUSANDS OF YEARS OF HISTORY TO CREATE AN ENCHANTING PLACE.

147 BOTTOM SWANS, WHICH NEST ALONG THE COAST, ARE UNMISTAKABLE BECAUSE OF THEIR SIZE AND PLUMAGE.

The Algarve
Portugal

THE ALGARVE - PORTUGAL - THE ALGARVE - PORTUGAL - TH

The word Algarve comes from the Arabic *al-gharb*, which literally means "to the west." The Algarve region lies in the southwest of Portugal, open to the Atlantic Ocean. Thanks to the presence of enchanting beaches and luxuriant gulfs, it is considered the country's most renowned and exclusive resort area. The Algarve is a land of contradictions. On one hand, entrepreneurs promote mass tourism; on the other, environmentalists are doing their best to protect and safeguard magnificent natural resources, for example, creating protected areas like the Rio Formosa Natural Park, which includes the estuary of the river of the same name. This park is a great attraction for wildlife enthusiasts. As often happens, when dealing with an estuarine zone, the biological diversity is significant, above all because two ecosystems (one of fresh water and the other marine) come together and create one particularly intricate and diversified ecotonal zone (i.e., a zone of "transition between different biological communities"). The Rio Formosa estuary itself is a series of dunes, canals, and salt marshes. Immediately behind the dunes extend vast expanses with lagoons and small islands of sand. During the alternation of the tides, the salty habitats are revealed; with the ebbing of the tide vast salt-water marshes appear. Given that it offers such different conditions over the course of the day, it is no surprise that the estuary is a place of great biodiversity, in part because of the immense quantity of food that is available and the variety in the degrees of salinity. All of these factors make the Rio Formosa estuary and the Algarve one of the most important of the wetlands between Europe and sub-Saharan Africa. Among the numerous bird species typical of estuarine areas, mention should be made of the purple gallinule, together with such other noteworthy birds as the grey plover, numerous limicolous species (those that prefer muddy environments), the kingfisher, ducks, and various birds of prey. As for fauna, the Algarve is an important safe habitat for the Iberian lynx, once much more widely distributed throughout the peninsula, but which now finds refuge and protection here. Among the dunes and pine forests, one of Europe's least-known reptiles, the common chameleon, moves at its characteristic slow pace. For this saurian, which was most likely imported from North Africa long ago, the Algarve one of its few bastions on European territory.

148 PONTA DA PIEDADE, IN THE VICINITY OF THE CITY OF LAGOS, IS CHARACTERIZED BY FASCINATING STACKS, SHAPED BY WIND AND WATER.

148-149 TO THE EAST OF THE ALGARVE COSAT IS THE CHARMING PORT OF SAGRES. HERE HISTORY AND LEGEND MINGLE, AND THE PRESENCE OF PRINCE HENRY NAVIGATOR LINGERS WITHIN THE WALLS OF THE FORTALEZA.

149 TOP MANY PEREGRINE FALCONS, RAVENS, AND MANY OTHER SEA BIRDS NEST ON THE TALL CLIFFS. FOR THIS REASON, THE ALGARVE IS OFFICIALLY LISTED AS AN IMPORTANT BIRD AREA (IBA).

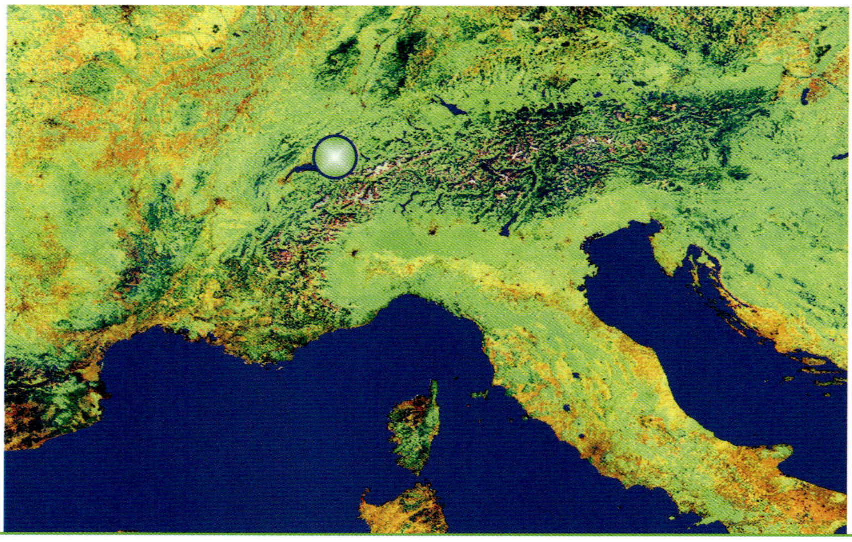

THE JUNGFRAU - SWITZERLAND - THE JUNGFRAU - SWIT

150 TOP AND 150-151 THE JUNGFRAU IS THE SECOND HIGHEST PEAK IN THE BERNESE OBERLAND ALPS; ONLY THE FINSTERRAARHORN OUTRANKS IT.

151 THE ALETSCH GLACIER IS A TRUE HIGHWAY OF ICE AND SNOW, STRETCHING UNINTERRUPTEDLY FOR ABOUT 14 MILES (22.5 KM).

152-153 THE ATMOSPHERE AT DUSK EMPHASIZES THE MAGNIFICENCE OF THE PEAKS OF THE EIGER (13,025 FT/3970 M), THE MÖNCH (13,449 FT/4099 M)), AND THE JUNGFRAU (13,642 FT/4158 M)).

The Jungfrau
Switzerland

Upon coming to the Bernese Oberland, visitors can only admire the breathtaking panorama of snowy peaks. Among them one in particular stands out, actually formed by three peaks rising over 13,000 ft (3962 m) from the plain. This mountain group is dominated by the Jungfrau, a soaring mass in which nature displays one of its most magnificent and inspiring faces. The two spires flanking the 13,642 ft (4158 m) mountain are famous: the Eiger (13,025 ft/3970) and the Mönch (13,449 ft/4099 m) recall expeditions that made mountaineering history. In geological terms, the Jungfrau is formed of overlapping layers of Jurassic limestone and gneiss, while its upper part is composed of gneiss alone. With the Eiger and the Mönch, the Jungfrau is the cradle of the mighty Aletsch Glacier, which extends for over 15 miles (24 km), a length that is unrivaled in the entire Alpine region.

Since 13 December 2001, the territory that is home to the immense glacier, along with the regions of Bietschhorn (12,907 ft/3934 m) and the Jungfrau area itself, has been included in UNESCO's Heritage of Humanity List. In this way, the part of the Valais, located around Europe's largest glacier became not only the first site in the Alps to win this distinction, but now, as a protected region, can continue to attract new generations of enthusiastic nature-loving visitors with its special, often gorgeous, sights.

The first protective measures applied to the Jungfrau area date back to the 1930s, when the Aletsch Forest Reserve was created, and since 1983, the area has been included in Switzerland's list of natural landscapes, sites, and monuments of national importance. This site features evidence of the Alps' geological history, thanks to the crystalline massif of the Aar and the presence of indigenous sedimentary layers that are of great interest to scientists. The Reserve's geological and geomorphological diversity, altitudes that range from 2900 ft (883 m) up to 14,023 ft (4274 m), and strongly contrasting climatic conditions (from the humid and cool oceanic climate to the hot and dry conditions in the interior valleys) are at the origins of the region's great biological wealth. It is home to many rare animal and plant species. Some of these are in danger of extinction, for example, the bearded vulture.

Perfect symbiosis between man and nature could give new inspiration for the conservation of natural environments by taking into particular consideration factors such as the landscape in its totality. It is a symbiosis that is urgently needed

The Jungfrau

154 TOP MARMOTS AND IBEXES ARE THE MOST TYPICAL REPRESENTATIVES OF THE FAUNA OF THE JUNGFRAU AND THE BERNESE OBERLAND.

154-155 AND 155 FROM KLEINE SCHEIDEGG, THE PEAKS OF THE JUNGFRAU AND THE SILBERHORN CAN BE SEEN. THE WHOLE AREA HAS BEEN RENOWNED SINCE THE EIGHTEENTH CENTURY FOR ITS SHEER ROCK WALLS, A FAVORITE PLACE FOR AN ENTIRE GENERATION OF MOUNTAINEERS AND CLIMBERS.

156-157 THE MÖNCH, TOGETHER WITH THE PEAKS OF THE EIGER AND THE JUNGFRAU, IS A COVETED DESTINATION FOR CLIMBERS, WITH COUNTLESS CLIMBING ROUTES OF GENERALLY HIGH DIFFICULTY.

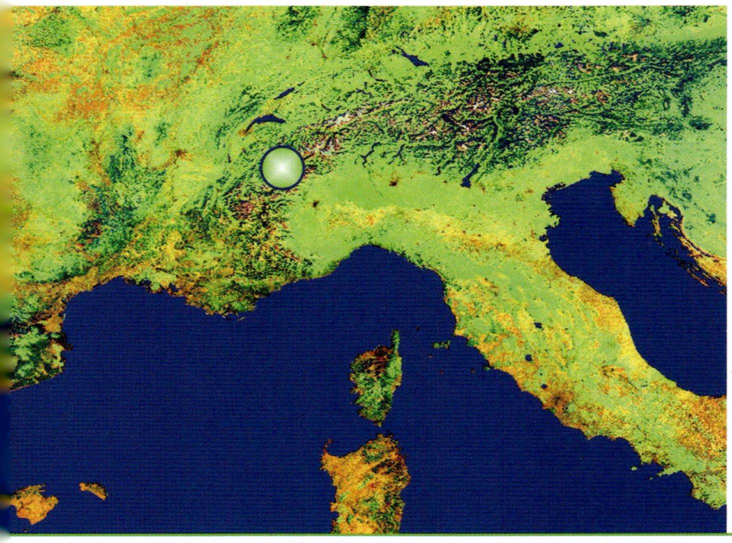

158 TOP MONT BLANC, OF GRANITIC ORIGIN, IS DOTTED BY SPIRES, TALL PEAKS, AND DEEP VALLEYS IN WHICH NUMEROUS GLACIERS HAVE DEVELOPED.

158-159 AND 159 MONT BLANC'S SUMMIT IS COMPLETELY COVERED BY A MANTLE OF ICE AND SNOW FROM 52 TO 75 FT (16 TO 23 M) THICK.

Mont Blanc

France/Italy/Switzerland

Mount Blanc, a geographical crossroads between France, Switzerland, and Italy, can only be definec as the king of the Alps. It is the tallest peak in Europe: to find any higher crest means a trip to the distinct Caucasus. Mont Blanc's vast summit, which registers an average temperature of ca. 1° Fahrenheit, rises to 12,337 ft (3760 m) above Chamonix, to 14,551 ft (4435 m) above Lake Geneva and finally, to 15,772 ft (4807 m) above sea level. From such a privileged viewpoint, one can see only a confusingly small portion of the globe, from the Apennines to the Vosges. This granite colossus surrounded by glaciers incorporates a series of imposing peaks, like the Dent du Géant and the Grandes Jorasses, which, isolated above enormous glaciers, offers hiking and climbing routes that are among the most difficult and demanding in the Alps. First climbed in 1786, for centuries Mont Blanc has constituted a barrier between Italy, France, and Switzerland, passable only by way of the passes of Col de la Seigne and Col Ferret to the southwest and northeast of the massif. Today, it is no longer like this: the group is crossable by road through a tunnel about 7 miles (11 km) long, and it is also served by a vast system of cable cars. From any viewpoint – naturalistic, panoramic, or cultural – Mont Blanc is an exceptional and unique entity, both for Europe and the world. Though great peaks outside Europe have been included in protected areas or established as national parks (e.g., Everest in Asia, McKinley in North America, Kilimanjaro in Kenya, the Ruwenzori Mts. in Africa, and Mt. Jaya in Oceania), Mont Blanc dominates no true protected area; no nature preserve has been created for what is not only the "roof of Europe," but also one of the most important monuments on the European Continent. The idea to include it within a park or a protected area was proposed several years ago, as the result of a few clear-cut considerations. From a naturalistic point of view, Mont Blanc encompasses some of the most typical ecosystems and species in the entire Alpine arc. It is of great interest in connection with alpine historical traditions and the value of its natural settings and environments. The massif's fauna is abundant and diversified and include almost all the species typical of the Alpine arc, and the flora, as well as being replete with expected local species, also contains many rarities. The geomorphological phenomena are spectacular, just as there are also precious examples of an alpine landscape shaped by the traditional activities of man: clearings in the woods and high-altitude pasturelands with area-specific plants. Finally, in addition to the values symbolized by Western Europe's highest peaks, the Mont Blanc massif includes some of the most important and magnificent unspoiled natural environments still to be found within the continent.

160-161 AMONG MONT BLANC'S MANY GLACIERS, THE MOST IMPORTANT ARE THE FRENEY, THE BRENVA, THE MIAGE, THE BREUILLAT, THE BOSSONS, AND THE MER DE GLACE (SEA OF ICE).

161 TOP VAL FERRET, TOGETHER WITH NEARBY VAL VENY, CONSTITUTES ONE OF THE MAIN TOURIST ATTRACTIONS IN COURMAYEUR BOTH IN WINTER AND IN SUMMER.

161 BOTTOM THE AIGUILLE NOIRE DE PEUTEREY IS THE MOST FAMOUS PEAK IN THE WHOLE VAL VENY, REACHABLE ONLY BY A VERY CHALLENGING PATH WITH FIXED IRON CABLES.

Mont Blanc

162 TOP LEFT THE ASCENT OF THE AIGUILLE DE MIDI ON SKIS IS AN EXCEPTIONAL CHALLENGE FOR HIGH-MOUNTAIN SKIERS. THE ROUTE FOLLOWS THE NORTHERN PART OF THE FRENCH SIDE OF MONT BLANC.

162 TOP RIGHT MONT BLANC'S HARSH ENVIRONMENT HAS MADE THIS MOUNTAINOUS MASSIF A TRUE GLACIAL PARADISE, WHICH HAS NO EQUAL AMONG THE OTHER PEAKS OF THE ALPS. THE CRESTS AND INCREDIBLE SHAPES FORMED BY THE GRANITE PEAKS GIVE RISE TO ECSTATIC REACTIONS AND POWERFUL EMOTIONS.

162-163 AND 163 THE FRENCH SIDE OF MONT BLANC LOOMS MAJESTICALLY OVER THE CITY OF CHAMONIX, A RENOWNED SKI RESORT.

164-165 THE DENT DU GÉANT, ON THE BORDER BETWEEN ITALY AND FRANCE, BETWEEN THE COL DU GÉANT AND THE GRANDES JORASSES, IS 13,170 FT (4014 M) IN HEIGHT.

165 MONT BLANC AND ITS SETTING MERIT SPECIAL RESPECT: NOTHING IN THE ENTIRE ALPINE ARC RIVALS ITS MAJESTIC AND INCOMPARABLE SCENERY NOR THE GREAT EPISODES OF MOUNTAINEERING HISTORY IT HAS OCCASIONED.

166-167 THE ROCKY CRESTS OF MONT BLANC BLEND INTO THE DENSE CLOUDS, TAKING ON SHADES MADE EVEN MORE MAGNIFICENT BY THE LIGHT OF DAWN.

The National Park of Gran Paradiso and the Vanoise Italy/France

Peaks and ibex, marmots, roe deer, and shepherds: these are all elements that appear as simple details in what is the largest protected area in Western Europe, Gran Paradiso-Vanoise. The two parks together total about 173,000 acres (70,000 hectares), an area equal to about half of Yosemite National Park in California. In placing this alpine territory under protection, Italy and France have secured a mountain territory of great naturalistic and scenic significance, a far-sighted action that also took into consideration the reality of the human settlements and the particular history of this area of the Alps.

The mountains forming the Gran Paradiso-Vanoise complex were forged by huge glaciers and rivers, which finally created wide valleys with typical alpine features. The area's history is very significant as concerns the realization of conservationist aims. In fact, we have to thank the kings of Italy for the foundation of Gran Paradiso Park. In the 19th century, members of the House of Savoy put the brakes on the hunting of the ibex, which was causing the population of this fascinating alpine inhabitant to diminish rapidly. As the traditional hunting practices of the past evolved toward unregulated and destructive ones, using ever more accurate and powerful firearms, this hoofed creature was quickly approaching extinction, to the point of numbering, in Italy, no more than 100 specimens. Consequently, however, thanks to the transformation of the area from a hunting reserve to a national park, the ibex was able to slowly repopulate the mountains, and now totals about 1500 in number. During the last 30 years or so, ibex from Gran Paradiso have been introduced into several other alpine zones, thus contributing to the population's recovery across the entire mountainous arc.

In terms of fauna, the Gran Paradiso-Vanoise complex has an unrivaled record in the conservation of the great alpine herbivores. Not only the ibex but also the chamois thrives; though now numerous, the chamois is more difficult to observe than the ibex due to its more withdrawn lifestyle. Not all the animal species have been as fortunate as the ibex: unfortunately, the wolf, the lynx, and the capercailzie, still present up to a century ago, have disappeared. However, the bearded vulture, the great bird thought to be extinct at the beginning of 20th century, is actually returning to the Alps thanks to an international reintroduction project promoted by Gran Paradiso Park, as is the lynx, that fascinating predator, which for the past few years has been timidly re-entering the protected area.

168 AND 168-169 THE MASSIF OF GRAN PARADISO IS FORMED OF A LONG WALL THAT RISES FROM THE PLAIN OF THE GLACIER OF TRIBULATION, A PROTECTED AREA ESTABLISHED IN 1922 THAT INCLUDES PART OF THE TERRITORY OF COGNE.

169 TOP THE GRAN PARADISO GROUP IS COMPOSED OF ROCKS OF VARYING AGES AND ORIGINS. IT IS FORMED OF A COMBINATION OF STRATIFIED GNEISS, METAMORPHIC ROCK DERIVED FROM GRANITE, AND DIORITE.

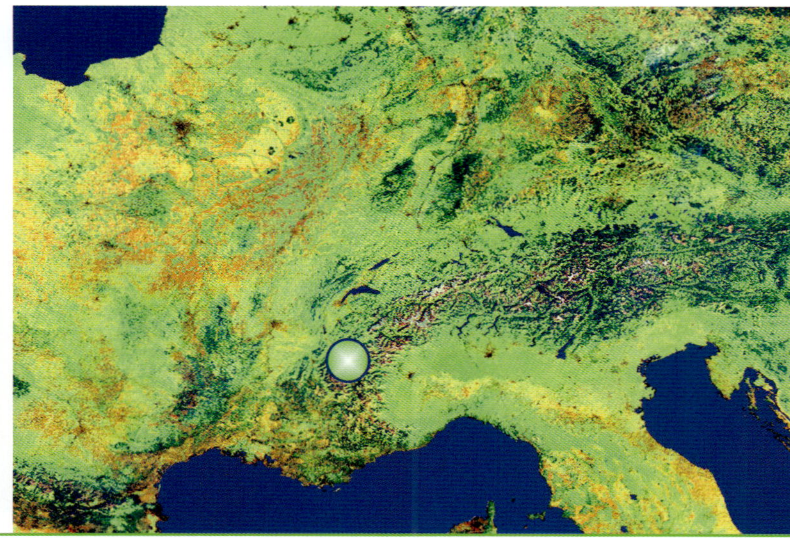

170 TOP LEFT THE BEST-KNOWN MAMMAL IN THE PARK IS THE IBEX. THIS MAJESTIC HOOFED CREATURE, NOW VERY NUMEROUS IN THE AREA, ONCE RISKED EXTINCTION, HAVING BEEN REDUCED TO ONLY ABOUT NINETY SPECIMENS.

170 TOP RIGHT AND 170-171 THE VANOISE MASSIF REPRESENTS THE NATURAL CONTINUATION OF THE GRAN PARADISO ON FRENCH TERRITORY. TOGETHER, THE TWO MOUNTAINS FORM THE LARGEST PROTECTED AREA IN EUROPE.

171 GRAN PARADISO, WHICH IS THE TALLEST PEAK IN THE GRAIAN ALPS, DIVIDES THE PIEDMONT FROM THE VAL D'AOSTA. FROM ITS PEAKS, ALL OF THE GRAIAN ALPS AND A GOOD PART OF THE REST OF THE ALPINE ARC CAN BE VIEWED.

Gran Paradiso and

the Vanoise

172-173 AND 173 TOP THE SERRÙ IS ONE OF GRAN PARADISO'S MOST FAMOUS LAKES. THE ROAD UP TO COL DU NIVOLET BRINGS SO MANY VISITORS THAT THEY CONSTITUTE AN ENVIRONMENTAL TREAT.

173 CENTER THE ABRUPT CHANGES IN THE PHISICAL SURFACE OF GRAN PARADISO GIVE RISE TO SPECTACULAR WATERFALLS, PARTICULARLY IN VALNONTEY.

173 BOTTOM ON THE SLOPES WITH GREATER SUN EXPOSURE, THE SNOW LINGERS FOR LESS TIME AND MAKES IT POSSIBLE FOR IBEXES, THE TRUE, UNCHALLENGED KINGS OF THE MOUNTAIN, TO GRAZE UNDISTURBED.

174-175 THE VILLAGE OF MONAL, IN THE VANOISE, OFFERS A GLORIOUS VIEW OF MONT POURRI.

175 TOP FOXES ARE NOT TYPICAL MOUNTAIN ANIMALS, BUT THEIR ABILITY TO ADAPT ALLOWS THEM TO TAKE ADVANTAGE OF THE GREAT AVAILABILITY OF NOURISHMENT ON GRAN PARADISO.

175 CENTER DURING SUMMER, THE LOWLANDS OF THE PARK ARE OFTEN DECORATED BY THE SUDDEN AND WIDESPREAD BLOSSOMING OF FLOWERS.

175 BOTTOM AMONG THE GRAN PARADISO AND VANOISE PARKS' MOST SIGNIFICANT NATURAL ELEMENTS ARE THE TORRENTIAL WATERWAYS THAT PROMOTE GROWTH OF VAST EVERGREEN FORESTS.

The Dolomites
Italy

THE DOLOMITES - ITALY - THE DOLOMITES - ITALY - THE

For many mountain enthusiasts and hikers of any nationality, the Dolomites are simply the most beautiful mountains in the world. Without doubt, if they cannot take *tout-court* first place for beauty, they are certainly among the most unusual peaks for landscapes and nature. The Dolomites rise suddenly toward the sky, reaching challenging altitudes (some above 9850 ft/3000 m), taking on soft pastel shades at dusk. Rightfully, the French architect Le Corbusier defined them as the most fascinating architectural work in the world: water and wind have shaped these mountains in a peculiar way, making them into natural pyramids or Gothic cathedrals. The Dolomites get their name from dolomite, the particular kind of calcareous rock of which they are composed. This rock, in turn, owes its name to the French geologist Déodat de Dolomieu, who "discovered" its chemical composition (calcium carbonate with magnesium, which mineral is called "dolomite," plus calcite in varying percentages), which makes this rock so "different" from the rest of the mineral kingdom. Of particular interest to understanding the Dolomites is their origins. These date back to the era in which the Tethys Ocean, a warm and tropical-type sea, in whose waters coral reefs developed, covered what is today northern Italy. The process is just like what happens today in tropical seas around the world: microscopic coelenterates created atolls and reefs, depositing layer upon layer of calcium carbonate. When the African continental plate collided with the European one, the Alps lifted along with the Dolomites, which are an integral part of them. The calcareous coral deposits that compose them were raised and have since then been exposed to erosion, which contributed among other things to facilitating the development of vast karstic phenomena including caves, galleries, waterways and other features.

The natural borders of the Dolomites region are situated within a polygon about 46 miles (74 km) long on each side, north of the Austrian border, east of the Adige River valley, south of the imaginary line between Belluno and Trento, and west of the Piave River valley. The Dolomites are distinguished by their geology and landscapes, but their flora and fauna, which are abundant and diversified, should not be overlooked. The fauna is typical of the eastern Alpine arc. Among the hoofed species, chamois, roe deer, and deer stand out, and the last-mentioned are currently growing in numbers. In the Belluno Dolomites, there is a substantial population of mouflons, an animal not originally found in the Alps and whose introduction to the region from Sardinia dates back to 1971. The Dolomites are also home to various *mustelidi*, among them ermines, martens, beech martens, and polecats, in addition to many rodents, lagomorphs, and insectivores. Large-size predators are still absent; they were intensively hunted in past centuries. The range is also renowned for its flora, with over 1500 plant species that paint the pasturelands and woods. Some 50 or so of the 120 species of orchids that grow in Italy can be found in the Dolomites. The species that grows at the highest altitude is the *Chamorchis alpina* the largest and, overall, most fascinating, is the yellow lady's slipper (*Cypripedium calceolus*), which often grows along the edges of moraines and landslides. Today many areas in the Dolomites are protected, thus contributing to safeguarding and protecting zones of natural importance. There are five protected areas in the province of Bolzano (Sciliar/Schlern, Odle/Geisler, Fànes-Sennes-Bràies, Sextner Dolomites and Tessa Group); three in the province of Belluno (Belluno Dolomites, Ampezzo Dolomites, and Tovanella, Somadida and Consiglio), three in the province of Trento (Paneveggio-Pale di San Martino, Stelvio and Adamello-Brenta), and two in Friuli (Friulian Dolomites and Prescudin).

176 BOTTOM AND 177 TOP RIGHT THE STRATIFICATION IN THE SPIRES OF THE BRENTA DOLOMITES SHOW SIGNS OF BIOTIC ORIGIN, WITH EROSION BY MARINE COELENTERATES.

176-177 AND 177 TOP LEFT THE SHADE OF COLOR THAT THE DOLOMITES TAKE ON AT DUSK AND DAWN IS CALLED ENROSADIRA.

178-179 CIMA TOSA IS THE TALLEST PEAK IN THE BRENTA DOLOMITES.

180 TOP SASSOLUNGO IS CONSIDERED THE SYMBOL OF THE VAL GARDENA.

180 CENTER AND 180-181 THE CATINACCIO GROUP IS FORMED OF THE TORRE VAJOLET, THE CATINACCIO D'ANTERMOIA, CIMA CATINACCIO, CRODA DI RE LAURINO, CRODA DAVOI, TORRE GARDECCIA, THE COGOLO DI LARSEC, THE CREPE DI LAUSA, CIMA SCALIERET, CIMA DELL POPPE, THE MASARÈ, AND THE MOLIGNON.

180 BOTTOM VAL GARDENA AND VAL BADIA ARE LINKED BY THE GARDENA PASS.

The Dolomites

182 TOP TO MOUNTAINEERING ENTHUSIASTS, THE THREE PEAKS OF LAVAREDO ARE THE SYMBOL OF THE ITALIAN DOLOMITES. THEY RISE COMPACT, HARMONIOUSLY ALIGNED, AND GORGEOUS WITH THEIR SHAPES AND COLORS ABOVE VAL DI SESTO.

182-183 DESPITE THE PHRASE "THE THREE PEAKS," LAVAREDO ACTUALLY HAS SIX: CIMA OVEST DI LAVAREDO, CIMA GRANDE DI LAVAREDO, CIMA PICCOLA DI LAVAREDO, PUNTA DI FRIDA, CIMA PICCOLISSIMA DI LAVAREDO, AND TORRE MINOR.

The Dolomites

184 TOP THE PO RIVER DELTA IS FORMED OF A MAZE OF CANALS AND LAND RECLAIMED FROM THE CLUTCHES OF THE WATER, THANKS TO A THOUSAND YEARS OF DRAINAGE PROGRAMS.

184-185 THE BANKS OF THE PO DELTA ARE DOTTED WITH SMALL WOODEN PILE DWELLINGS USED BY FISHERMEN.

185 BOTTOM THE PO DELTA IS COMPOSED OF A MAZE OF CANALS AND LANDS RECLAIMED FROM THE RIVER THROUGH THOUSANDS OF YEARS OF DRAINAGE WORK.

The Po River Delta

Italy

The Po Delta is a region of numerous ecosystems; it has countless species of plants and animals and an atmosphere rich in tradition. The delta's abundant wildlife is nourished by its biological diversity, together with the remarkable ecological heterogeneity of the area. As a result, the delta is an European refuge for many birds, including some particularly interesting species. The only colony of the small cormorant is here, as are the only nesting pairs of the lesser-crested tern to be found in Europe. Here too one of the few European colonies of the flamingo is found. To ensure the health and maintenance of the delta, two regional parks have been founded to protect it. The one in the Romagna region was founded in 1988; the one in the Veneto region was founded in 1997.

Like the Camargue and the Danube Delta, the Po Delta is a meeting place of nature and tradition: the vast natural or re-naturalized areas alternate with places where the human presence and history are significant. It is therefore obvious that governing the area has to take into careful consideration both conservation and development factors. In the delta are plants adapted to living on beaches and dunes or wetlands; whether they are on lagoons and brackish depressions or marshes and fresh-water wetlands, they are vulnerable to any ill-considered development projects.

Despite human settlement being often massive and significant, the Po Delta still offers riverbeds whose courses are marked by abundant twists and turns and islands; the outfalls of streams, open shores with a rolling alternation of dunes, lagoons and fresh-water marshes; as well as ancient build-ups like sandbars with fossilized dunes and ancient riverbeds (*paleoalveo*). All of these environments feature interesting plant relationships, now the subject of intensive research and conservation efforts. In addition there are a few surviving ancient woods and some coastal pine forests within the region The fauna of the Po Delta Park is one of its most valuable assets, and more than 460 vertebrate species have been recorded. Bird life in particular is of extraordinary value; in the last few decades over 300 species have been catalogued, of which at least 150 nest and 180 spend the winters in the delta. Several species, both migratory and non-migratory, are particularly noteworthy because of their rarity. In all, such abundance makes the park Italy's most important ornithological area, and also one of the most important in Europe.

The Po River Delta

186 STARTING IN MAY, THE LESSER BITTERN STARTS NESTING AMONG THE DENSE CANE-FIELDS THAT LINE THE CANALS OF THE PO RIVER DELTA.

186-187 AMONG MALLARD DUCKS, MALES ARE CHARACTERIZED BY THEIR VERY BRIGHT COLORED PLUMAGE. THEIR HEAD IS DARK GREEN, THEIR CHEST REDDISH, AND THEIR BEAK YELLOW.

187 TOP LEFT TWO COCK PHEASANTS FIGHT DURING THE MATING SEASON.

187 TOP CENTER THE SQUACCO HERON PREFERS TO LIVE IN MARSHY AREAS, WHERE IT CAN FEED ON SMALL FISH, AMPHIBIANS, AND AQUATIC INVERTEBRATES.

187 TOP RIGHT GREY HERONS ARE ENDOWED WITH A WIDE WING SPAN, WHICH IN ADULT MALES MAY EXTEND TO OVER 6 FT (1.8 M).

188 TOP LEFT THE PIANO DEI PANTANI PLAINS ARE OCCUPIED BY A FEW PONDS THAT IN ANCIENT TIMES FORMED A LAKE.

188 TOP RIGHT THE ABUNDANT CALCAREOUS ELEMENT IN THE SIBYLLINE MOUNTAINS HAS RESULTED IN MANY CAVES, GALLERIES, SINKHOLES AND OTHER KARSTIC PHENOMENA.

188-189 ABOVE 6000 FT (1830 M), BEECH TREES BECOME THE DOMINANT PLANT SPECIES.

190-191 IN LATE SPRING, THE GREAT PLAIN LIGHTS UP WITH ALL THE COLORS OF THE MOUNTAIN FLOWERS.

The National Park of the Sibylline Mountains

Italy

Some places in the world are wrapped in an aura of mystery, which without doubt adds greatly to their interest. The Sibylline Mountains, for example, since 1933 protected within a national park, are cloaked in legends. One of these has to do with the presence of the famous Sibyl (from whom the mountain chain gets its name). Unfortunately, the cave in which she was supposed to have dwelled was closed off by a landslide a few years ago. On the other hand, another tale relates to Pilate's Lake, in which legend has it that the body of the Roman procurator involved in Christ's trial was laid to rest.

In reality, however, the Sibylline Mountains are of special appeal for a series of natural and environmental characteristics that make the area one of the most interesting protected zones in all of the Apennines. This chain stretches down central Italy, with the record for height held by Mt. Vettore, which reaches 8124 ft (2746 m) and owes its origins to the long process of sedimentation in the ancient Mediterranean Sea and the subsequent orogenesis or "birth" of the range. Then, in the Quaternary Period (from 1.8 million years ago to today), the Apennine glaciers caused a deep and intensive process of erosion, still apparent in a large number of U-shaped terminal valleys, such as those on Mt. Vettore, Mt. Bove, the upper Ambro Valley, and the Val di Tela. The Sibylline range's calcareous component has also resulted in enormous karstic formations, with many caves and dolines (sinkholes). Among the most interesting formations are the springs of Norcia, able to supply the local Apennine grasslands, known as the *marcite* (water-meadows), with water throughout the year, even in winter.

The heart of the massif and one of the most highly visited sections of the park is the crown of peaks that includes the Cima del Lago (7947 ft/2422 m), the Pizzo del Diavolo (7907 ft/2410 m), the Cima del Redentore (8032 ft/2448 m), and Mt. Vettore (8124 ft/2746m). The Cima del Lago encompasses Pilate's Lake, onto which faces the eastern wall of the Pizzo del Diavolo.

Among the Sibylline Mountains' most interesting wildlife features is the presence of a small aquatic crustacean unique in the world, the Marchesoni chirocephalus (*Chirocephalus marchesonii*), a characteristically red-colored shrimp-like creature that lives in the waters of Pilate's Lake, at 6368 ft (1940 m). The species, discovered in 1954 by Professor Marchesoni, then director of the Botanical Institute of the University of Camerino, survives only in this lake (which, during times of drought, tends to dry up, creating two connected ponds) and features an unexpected phylogenetic and biogeographical relationship to *taxa* (systematic categories into which the various living species are classified) native to the Asian continent. However, the entire Pilate's Lake area is of great environmental importance, featuring varieties of high-altitude vegetation that grow all along the moraines. The varieties include mountain avens, edelweiss, and pirin poppies.

Naturally, the rich animal life of the Sibyllines is not limited to the chirocephalus. Birdlife is very abundant and boasts species typical to the Apennines, including many birds of prey. Among the mammals are wolves, roe deer, porcupines, and wild cats, in addition to many rodents and insectivores. The reptile population includes Orsini's viper, which in central Italy is found only on a few other massifs, such as Gran Sasso in the Abruzzi. Among the invertebrates, 700 species of *Lepidoptera* (butterflies and moths) have been catalogued as of today, not to mention an endemic beetle (*Duvalius ruffoi*), dependent on Mt. Vettore for its survival.

The vegetation of the Sibyllines is typical of the Apennines. Up to about 3000 ft (915 m), mixed deciduous forest is prevalent. Above this height beech woods dominate up to about 6000 feet (1830 m), above which height stretch meadows with dwarf rhododendron bushes.

192 TOP LEFT IN SPRING, THE SILHOUETTE OF THE STILL-SNOWBOUND APENNINES CONTRASTS WITH THE COLORS OF THE FRESHLY BLOOMING FLOWERS ON THE GREAT PLAIN.

192 TOP RIGHT AND 192-193 PILATE'S LAKE, ONE OF THE MOST INTERESTING IN THE WHOLE SIBYLLINE MOUNTAIN AREA, IS HOME TO THE WORLD'S ONLY KNOWN POPULATION OF CHIROCEPHALUS MARCHESONII. THIS SMALL, BRIGHT-RED AQUATIC CRUSTACEAN INHABITS THE WATERS OF THE LAKE DURING THE SUMMER, AND ITS EGGS SURVIVE THE HARSH WINTER, ALLOWING THE SPECIES TO PROPAGATE ITSELF EVERY YEAR.

193 THE "FOSSO DEI MERGANI," OR MERGANI TRENCH, ENDS IN A SWALLOW-HOLE, A HOLLOW THAT CHANNELS THE WATERS OF THE WHOLE CATCHMENT BASIN. THE WATERS THAT FEED THE TRENCH COME FROM MELTED SNOW AND RAIN.

194-195 KARSTIC PHENOMENA ARE COMMON THROUGHOUT UMBRIA IN PARTICULAR IN THE AREA OF THE FOSSO DEI MERGANI. THERE, THERE ARE EROSIVE VALLEYS AND DOLINES IN VARIOUS SHAPES: FUNNEL-SHAPED, FLAT, AND BOWL-SHAPED.

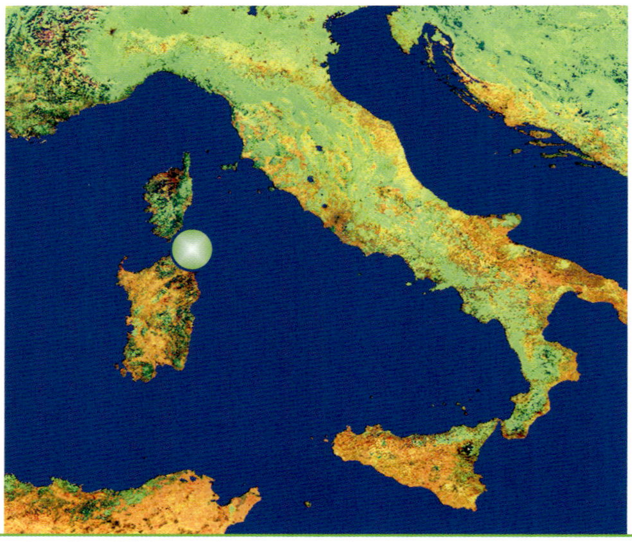

196 TOP THE MADDALENA ARCHIPELAGO IS COMPOSED OF A GROUP OF ISLANDS NORTHEAST OF SARDINIA, OFF THE EMERALD COAST.

196-197 AND 197 THE INDENTED COASTS OF THE MADDALENA ARCHIPELAGO ARE REFLECTED IN THE EMERALD BLUE OF THE SEA.

198-199 BUDELLI ISLAND'S PINK BEACH IS ONE OF THE MADDALENA ARCHIPELAGO'S BEST-KNOWN PLACES.

The National Park of the Maddalena Archipelago

Italy

The area between Sardinia and Corsica is very attractive because of its geological formation and the natural features that characterize it. Wedged into this tract of sea, often ruffled or whipped up by the *maestro* blowing from the northwest, the Maddalena Archipelago and the national park of the same name encompass about 60 islands and islets, noted for crystalline waters and enchanting green landscapes, characterized by unique biological relationships termed "biocoenoses," meaning communities of living creatures of different species that share the same space and compete within it for survival). Only the Mouth of Bonifacio strait separates the Maddalena Islands from the Lavezzi Islands, which belong to Corsica but are geographically related to the same archipelago and, from a geomorphological and landscape point of view, feature similar characteristics. The seven main Maddalena Islands, listed from largest to smallest, are: Maddalena, Caprera and Santo Stefano, Spargi, Budelli, Razzoli, and Santa Maria. Only Maddalena is inhabited; the other islands have no human settlements. For this reason, they still feature intact wild environments. In terms of plant species, many that appear in the archipelago are native or rare. Among the 700 floral species, *Silene velutina*, *Colchicum corsicum*, *Helicodyceros muscivorus*, a few *Limonia* are noteworthy, in addition to several species of major interest that are listed in the European Habitat Directive; these include *Cynomorium coccineum* and *Armeria pungens*.

Among reptiles native to the Sardinian-Corsican system are several important species, including two Lacertidae (lizards): *Archeolacerta bedriagae* and *Algyroides fitzingeri*. Other rare and important species now threatened, both on the local and worldwide level, include the gecko (*Euleptes europea*) and some land turtles. Birds are also very well represented, with substantial populations of Audouin's gulls (the only seagull species native to the Mediterranean) and the Mediterranean subspecies of the green cormorant. The populations of these two birds within the protected region represent respectively one and three percent of the world population. In addition, the population of Cory's shearwater (*Calonectris diomedea*), with about 1800 nesting pairs, constitutes three percent of the species' total Mediterranean population. The Maddalena Archipelago's most important species of marine invertebrate is the ribbed Mediterranean limpet (*Patella ferruginea*), a gastropod mollusk now restricted to a few small islands in the western basin of the Mediterranean and subject to strict protection laws. Because of its rarity, this limpet is included on the "Red List" of species threatened by extinction. On the islands of the Maddalena Archipelago, this mollusk is found in a few largely exposed environments, among them the promontories on the north and west of the islands of Spargi, Razzoli, Budelli, and Monaci.

The Maddalena Archipelago

200 THE STRAIT OF "CHIECCA DI MORTO" SEPARATES BUDELLI ISLAND FROM RAZZOLI AND SANTA MARIA ISLANDS, JUST TO THE NORTH.

201 TOP SPARGI ISLAND, PARTICULARLY ABUNDANT IN WATER, IS THE THIRD LARGEST ISLAND IN THE ARCHIPELAGO.

201 CENTER AND BOTTOM THE GRANITE CLIFFS THAT FORM CAPRERA'S NORTH AND EAST COASTS MAKE THE ISLAND ALMOST COMPLETELY INACCESSIBLE TO VISITORS.

202-203 THE MADDALENA ARCHIPELAGO IS FORMED OF BLOCKS OF GRANITE SHAPED BY WIND AND RAIN. THEY ARE NICKNAMED TAFONI.

The Lipari Islands

Italy

204 STROMBOLI IS FORMED BY A STILL ACTIVE VOLCANO, WHOSE ERUPTIONS SELDOM PRESENT A THREAT TO HUMAN SETTLEMENTS.

204-205 VULCANO ISLAND IS LOCATED IN THE SOUTHERNMOST PART OF THE LIPARI ARCHIPELAGO.

205 TOP LIPARI IS THE LARGEST ISLAND IN THE ARCHIPELAGO. THE ISLAND IS OF VOLCANIC ORIGIN AND ITS PRESENT-DAY APPEARANCE IS THE RESULT OF COUNTLESS ERUPTIONS AND LENGTHY DORMANT PERIODS.

The origin of the Lipari Islands is volcanic in nature. According to the most recent dating done by volcanologists, volcanic activity created the arc of the archipelago during the last 500,000 years. The islands making up this gorgeous necklet number seven: Lipari, Panarea, Vulcano, Stromboli, Alicudi, Filicudi, and Salina, and their morphology and genesis represent, among much else, an evolutionary model for the study of volcanology worldwide. The only volcanoes still active today are on Stromboli and Vulcano islands, but there is still thermal activity and fumaroles on Panarea and Lipari, whereas on Alicudi, Filicudi, and Salina internal volcanic activity has practically ceased.

From a biological point of view, the plant and animal populations on the Lipari group are the result of the natural colonization process of the islands, in which, during the last 7000 years, man has participated as well, introducing numerous plants and animals.

The archipelago is characterized by typical Mediterranean scrub, the result of thousands of years of interaction between man and nature: the most widespread plants and tree species on the islands are myrtles, rock roses, heather, and oleander. The landscape occupied by man (known as "anthropic") is dominated by olive trees and vineyards, not to mention the formation of scrub and pioneer plants. The slopes of the Liparian volcanoes are the preferred environment for specialized plant species: only a few pioneer shrubs have been able to take root, the most widespread of which is Tyrrhenian broom, a pulse vegetable endemic to the Lipari and the Ponziane Islands. This plant flowers between May and June, coloring the top of Stromboli a bright yellow.

In the migration season, many migratory birds pass through and rest on the islands. Among the nesting and non-migratory species are both Cory's and little shearwaters; among the few birds of prey are buzzards (*Buteo buteo*) and kestrels (*Falco tinnunculus*). The most interesting species of the falcon type is Eleonora's falcon (*Falco eleonorae*), which lives in colonies on the western cliffs of some islands in the archipelago.

Among the land vertebrates found on the islands, is a hard-to-observe subspecies of the Liparian garden dormouse (*Eliomys quercinus liparensis*) endemic to Lipari, though on some minor islands, small populations of the native Lipari lizard (*Podarcis raffoni*), darker than other species in the archipelago, survive. Studies conducted by herpetologists have demonstrated that this lizard was probably present on other islands of the archipelago in the past, but was eliminated through competition with the field lizard (*Podarsis sicula*), the other species that is found today in the archipelago. The Lipari Islands have no poisonous snakes, only the western whip snake (*Hierophis viridiflavus*), a harmless coluber. Other fairly common reptiles are the two geckos (*Tarentola mauritanica* and *Hemidactylus turcicus*), often seen on the walls of homes.

206 TOP THE VEGETATION ON LIPARI IS COMPOSED MAINLY OF MEDITERRANEAN PLANTS, LIKE OLEANDER, ROCK ROSE, AND BROOM AND OTHER AROMATIC TREES LIKE OLIVE TREES AND GRAPEVINES.

206-207 THE MEDITERRANEAN SCRUB THAT COVERS ALL OF THE ISLANDS IN THE ARCHIPELAGO CONTAINS SEVERAL ENDEMIC SPECIES OF GREAT VALUE, AMONG THEM CYTISUS AEOLICUS AND KOCHIA SAXICOLA.

207 TOP MT. FOSSA FELCI, AT 2539 FT (774 M), IS THE HIGHEST POINT ON FILICUDI ISLAND. THE ISLAND WAS KNOWN IN ANCIENT TIMES AS PHOENICUSA OR PHOENICODES (FULL OF FERNS), AN EXPLICIT REFERENCE TO THE DWARF PALMS STILL FOUND THERE.

207 BOTTOM PANAREA, ALONG WITH SOME OTHER ISLANDS (BASILUZZO, SPINAZZOLA, LISCA BIANCA, DATTILO, BOTTARO, LISCA NERA, THE PANARELLI, AND THE FORMICHE), CONSTITUTES A SMALL ARCHIPELAGO LOCATED BETWEEN LIPARI AND STROMBOLI.

208 VULCANO ISLAND HAS MANAGED TO KEEP ITS NATURAL ENVIRONMENT AND MORPHOLOGY ALMOST UNCHANGED. FOUR CRATERS, THE LARGEST OF WHICH IS FOSSO VOLCANO, ATTEST TO THE ISLAND'S VOLCANIC ORIGIN.

208-209 THE VOLCANIC CONE OF STROMBOLI HAS A VERY EVEN SHAPE, WITH VERY STEEP SLOPES. IT IS CHARACTERIZED BY ERUPTIVE ACTIVITY CALLED STROMBOLIAN, WITH MAGMA OF MEDIUM-LOW VISCOSITY AND LARGE QUANTITIES OF GAS.

Mount Etna

Italy

210 AND 210-211 THE BEAUTY OF ETNA DOES NOT DEPEND ONLY ON THE MAGNIFICENCE OF ITS ERUPTIONS, BUT ALSO ON THE UNIQUE AND INCOMPARABLE ATMOSPHERE THAT SURROUNDS IT, AS WELL AS THE GEOLOGICAL NATURE OF ITS CALDERAS.

211 TOP ETNA, THE HIGHEST ACTIVE VOLCANO IN EUROPE, IS OVER 25 MILES (40 KM) IN DIAMETER AT ITS BASE. BECAUSE OF ITS FREQUENT ERUPTIONS, WHICH LAST FROM A FEW DAYS TO SEVERAL MONTHS, THE MOUNTAIN IS ALWAYS IN A STATE OF CHANGE.

Etna, which is also known as Mongibello (from Mons Gebel, from the Latin *mons* and the Arabic *gebel*, both meaning "mountain"), is the highest active volcano in Europe, rising about 10,902 ft (3323 m). Since 1987, the Etna region has been protected within a nature reserve, founded to safeguard the wild life and the natural setting surrounding the volcano itself. In reality, the procedure for the creation of the protected area was very long and laborious, beginning of the 1960s when several leading environmentalists promoted the idea that Etna should be adequately protected, not least because of its unique status within the European continent.

It is difficult to describe the beauty of Etna fairly and in full. Though some aspects of its appearance are immediately attributable to lava eruptions, there are also tracts of great scenic and natural character and interest. In some areas, for example, there are fascinating and mysterious caves with lava running through them and underground passages that are fine summaries of geological and mineral origins, to say nothing of the area's true speleological biodiversity, which has equals in only a few parts of the world. From the biological perspective, it is interesting to observe how the fauna on Etna is still on the whole abundant. Unfortunately, here like elsewhere, large predators such as the wolf and the lynx became extinct; this occurred more or less concurrently with construction of the roads giving access to the heart of the area. Contrarily, the small- and medium-sized fauna is still abundant and diversified, proof of the area's great wildlife value. Among the land mammals are the marten, the porcupine, and many rodents and insect-eaters.

Etna can claim many bird species, in particular birds of prey, confirming the existence of vast uncontaminated spaces. The local vegetation also deserves mention: because of the alternation of zones of old lava flows with zones of more recent flows, it is possible to observe the whole range and stages of plant growth. In the areas of recent flows, for example, one can note first and foremost the presence and growth of crusty lichens, alternating with succulent species and plants particularly suited to open areas. Where the lava flows are but a memory, true woodlands have returned. On Etna, the beech trees have reached the southern limit of their area, thanks above all to the heterogeneity at that altitude and the presence of true microclimates. Even the birch on Etna, more often identified as a strictly endemic species, is present in areas of older flows. In areas of more recent flows, on the other hand, it is possible to observe heathlike astragalus plant formations, which offer shade to other plants, as well as many small-sized animals.

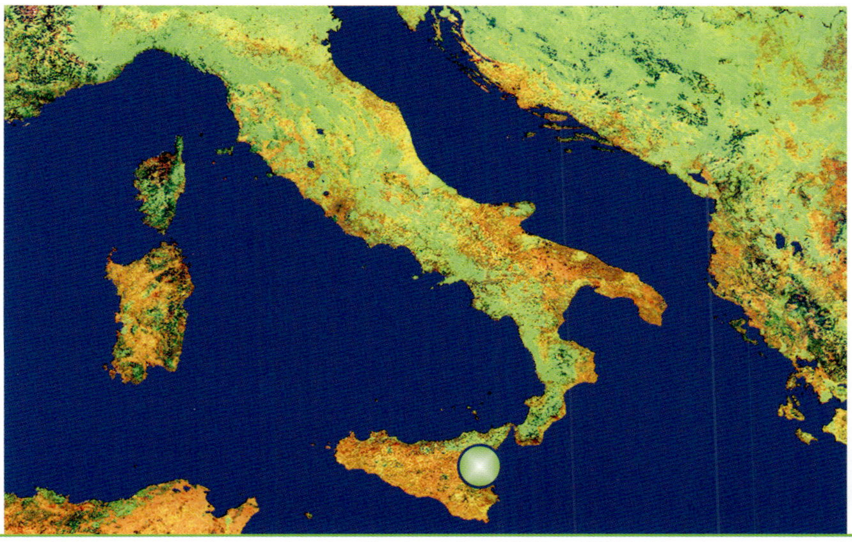

212 IN WINTER, AN ABUNDANT BLANKET OF SNOW COMPLETELY COVERS MOUNT ETNA.

212-213 AND 213 TOP ETNA REMAINS VERY ACTIVE, AS ATTESTED TO BY THE NUMEROUS PUFFS OF SMOKE THAT RISE BOTH FROM THE VOLCANO'S CRATER AND ITS SIDE CONES THROUGHOUT THE YEAR.

214 AT 10,902 FT (3323 M) IN HEIGHT, ETNA TOWERS ABOVE THE MEDITERRANEAN, RISING ABOVE ONE OF THE HOTTEST SEISMOGENETIC SYSTEMS IN ALL OF EUROPE. DURING ERUPTIONS, LAVA, THRUST OUT WITH FORCE, FORMS TRUE RIVERS THAT RUN DOWN ITS SIDES.

215 THE EXPLOSIVE ACTIVITY OF ETNA IS LIMITED TO STROMBOLIAN VOLCANIC PHASES AND ONLY RARELY TO PRODUCING FOUNTAINS OF LAVA. THE OVERALL ETNEAN ERUPTIVE PROCESS IS GEOLOGICALLY RECENT: ONLY 700,000 YEARS OF ACTIVITY, A PERIOD DURING WHICH THE VOLCANO'S ERUPTIONS HAVE PASSED THROUGH MANY PHASES, BOTH EFFUSIVE AND EXPLOSIVE.

216-217 A RIVER OF MAGMA FLOWS GLOWINGLY BETWEEN WALLS OF VOLCANIC ROCK. ONCE SOLIDIFIED, IT WILL BE COMPLETELY LACKING IN NITROGEN AND WILL THUS BE VERY HARD FOR PLANTS TO COLONIZE.

218 TOP THE ISLAND OF COMINO IS UNINHABITED AND, FOR THIS REASON, INTACT FROM AN ENVIRONMENTAL POINT OF VIEW.

218-219 AND 219 GOZO IS THE SECOND LARGEST ISLAND IN THE MALTESE ARCHIPELAGO AND IS CHARACTERIZED BY MANY SITES OF GREAT NATURAL BEAUTY, LIKE THE AREA OF DWEJRA POINT.

220-221 THE ISLAND OF COMINO, ARID AND ROCKY, LONG A HIDE-OUT FOR PIRATES, IS TODAY ONE OF THE PLACES MOST LOVED BY TOURISTS AND SCUBA DIVERS.

The Préspa Lakes

Greece

222 THE DIVERSITY OF THE BIRDS THAT LIVE AROUND THE PRÉSPA LAKES IS REMARKABLE, ABOVE ALL FOR LIMICOLOUS AND MARSHLAND SPECIES.

222-223 AND 223 TOP THE ENVIRONMENT SURROUNDING THE MIKRÌ PRÉSPA AND MAGALI PRÉSPA LAKES IS STILL WELL PRESERVED.

224-225 MIKRÌ PRÉSPA LAKE IS ONE OF THE MOST IMPORTANT WETLAND BIOTOPES IN ALL OF GREECE. IN PARTICULAR, IT STANDS OUT FOR ITS NOTEWORTHY PELICAN POPULATION.

The system used in Europe to identify areas that should be protected gives special consideration to wetlands, which, when receiving other protective measures, are often included in the Ramsar Convention or on the Important Bird Areas list. This is the case with the Mikrì Préspa National Park in Greece, on the border with Albania, which encompasses three fairly shallow lakes: Mikrì Préspa (Little Préspa), Lake Ohrid, and Megali Préspa (Great Préspa). The most interesting characteristic of this area of about 20,000 acres (8100 hectares) is the presence of muddy and flooded fields and a cane-field zone, the proper management of which has made possible the development of very varied bird life.

The abandonment of traditional forms of livestock raising, which favored the presence of wetlands, and the growth in the human population have unfortunately caused a progressive reduction of meadowland, with the concurrent excessive growth of the cane-fields. The carp population, once highly abundant, has been growing smaller, while many aquatic birds have declined in number or even disappeared. Currently, a series of initiatives aimed at re-stabilizing the lakes' original condition is being drawn up. The park's flora is equally abundant, with over 1200 catalogued species, and in particular, extensive surfaces covered with Phragmites and Typha. The fauna is typical of the geographical area, with about 30 species of mammals, among them the otter, the jackal, the wolf, and the brown bear. The nutria has been introduced here (as in a number of other locations in Europe) and is still a very destructive force as it feeds on plant materials and weakens the shores with the building of its dens. Furthermore, the nutria have damaged important cane-fields, which represent one of the main habitats for the reproduction of many bird species.

Of particular importance among the avian species are the colonies of the Dalmatian pelican (*Pelecanus crispus*) and the white pelican (*P. onocrotalus*), two species that are now very rare in Europe, found only in the Danube Delta and in some zones of the ex-Soviet Union. In 1984, the populations of these two species amounted to 165 and 116 pairs respectively, numbers that have remained stable over the years. On the other hand, the number of the *ardeidae* (herons and night herons) has varied; many species have demonstrated a noticeable decline over the years.

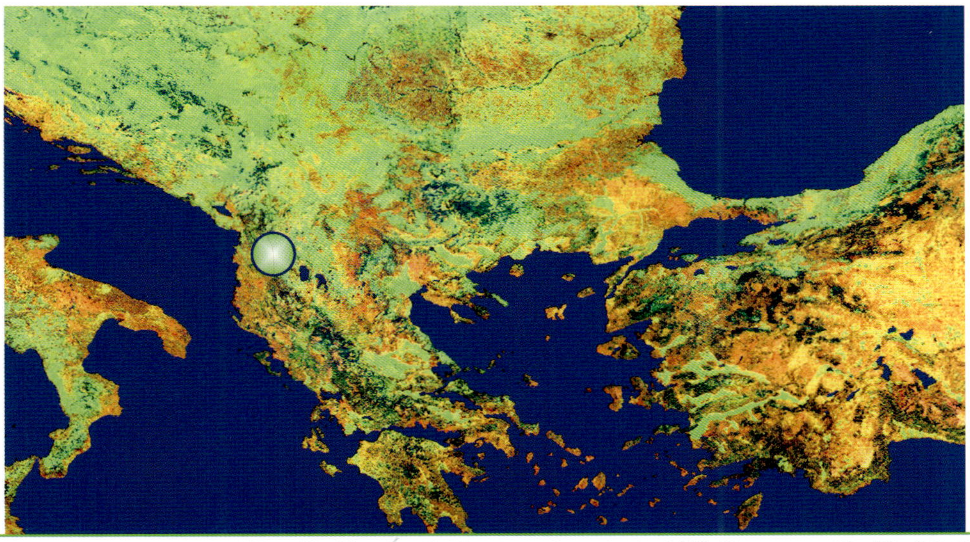

THE PRÉSPA LAKES - GREECE - THE PRÉSPA LAKES - GREECE -

The Ionian Islands
Greece

THE IONIAN ISLANDS - GREECE - THE IONIAN ISLANDS -

The Ionian Islands lie off the western coast of continental Greece. The archipelago is composed of seven large islands, lying on a northwest-to-southeast axis. They are (from the north): Corfu (Kerkira), Paxí, Lefkada, Kefaloniá, Itháki, Zákinthos (Zante), and Kíthira. The smaller islands include Meganissi, Antipapi, Antikythira, and Elafonissos.

The Ionian Islands rate high among Greece's most popular tourist destinations, with visitors attracted by the relatively low cost of living and highly appealing landscapes and towns. Tourist facilities and activities on the islands are generally concentrated on the eastern shores, with their gentle slopes and generous beaches; the western shores feature tall, rocky, much less accessible cliffs that are certainly no less beautiful – besides being the area preferred by bird species. One of the archipelago's most distinctive characteristics is its remarkable vegetation. Corfu, for example, is the most humid place in all Greece, with an average of over 8 inches (20.5 cm) of rain in December. This allows for the growth of many hydric plants (humidity lovers) and the presence of more typically northern species. The high humidity also guarantees the presence of many orchids, of which at least 36 species are found on Corfu. In general, besides being humid, Corfu and the other islands can be somewhat muggy, in particular when the *meltemi*, the dry, cool northwest wind, is not blowing.

One highly interesting element of the vegetation is the endemic fir tree, the Greek fir (*Abies cephalonica*), found on the islands and in continental Greece. On Kefaloniá, it is the dominant forest species on Mt. Ainos, where its purity has been maintained thanks to the absence of cross-breeding. The Greek fir grows between the 2600 ft (800 m) level and 5250 ft (1600 m) level, reaching a height of 100 ft (30 m). It forms dense forests on the Greek mountains and was in the past an important source of wood, prized for shipbuilding. Today, Mt. Ainos, which reaches the noteworthy altitude of 5341 ft (1628 m), is protected within a national park that specifically provides a home to the fir-tree forest. A number of very valued bird species are native to the archipelago. Some, like yellow wagtails, larks, thrushes, and sparrows, most often nest in the olive trees and the few remaining forest areas, whereas others, among them herons and lesser egrets, rest in the archipelago during their migration from Africa to Europe. Large mammals such as deer, wild boar, and roe deer, and also many marsh birds, are now extinct on the islands, above all because of excessive hunting. Today, only small mammals, like foxes, squirrels, porcupines, and bats, are relatively widespread. By contrast, herpetological species – the reptiles – are quite abundant, with many lizards and snakes. The islands are very significant to wildlife conservationists because they are home to one of the most important remaining colonies of monk seals, one of the rarest and most threatened mammals in European waters and one of the six mammals in the world that the International Union for the Conservation of Nature and Natural Resources (IUCN) has listed in its "critically endangered" category. In the waters around the archipelago are also other marine animals of great interest to conservationists, including dolphins and other odontocetes (toothed) whales.

The waters also have a substantial population of *Caretta caretta* – the sea turtle: on the island of Zante, the turtle's most important breeding beach, between 700 and 1000 nests are being recorded annually. Thanks to the presence of the Zante National Marine Park, Greece has become fully conscious of the environmental dangers that mass tourism brings. The park's main tasks are to protect the island's turtles and seals, as well as to safeguarding the marine ecosystems.

226 BOTTOM ZANTE, ALSO KNOWN AS ZAKYNTHOS, WAS, FOR A LONG PERIOD, A VENETIAN COLONY, NICKNAMED "FLOWER OF THE LEVANT" BECAUSE OF ITS BEAUTY.

226-227 ONE OF THE MOST POPULAR SPOTS ON ZANTE IS SHIPWRECK BEACH, A LITTLE SANDY BEACH WEDGED INTO A DEEP FJORD SURROUNDED BY SHEER WALLS.

227 TOP ZANTE, WITH OLIVE AND ALMOND GROVES, WOODED HILLS, AND SHEER COASTLINES, IS ALSO DOTTED BY SMALL AND QUIET BEACHES.

228 KEFALONIÁ BOASTS NUMEROUS KARSTIC SYSTEMS. ONE OF THESE IS MELISSANDI LAKE, WHICH FORMED AS A RESULT OF THE SINKING OF THE CAVE OF THE SAME NAME IN ANCIENT TIMES.

228-229 THE VILLAGE OF ASSOS ON KEFALONIÁ WAS FOUNDED ATOP A VAST PROMONTORY COVERED BY OLIVE AND CYPRESS TREES.

231 TOP CENTER PALEOKASTRITSA BAY IS ONE OF THE MOST RENOWNED PLACES ON CORFU.

231 TOP RIGHT LEFKADA ISLAND, WITH ITS ENDLESS BEACHES, IS LINKED TO MAINLAND GREECE BY A MOBILE BRIDGE.

230 AND 230-231 PAXI ISLAND IS ALMOST COMPLETELY SURROUNDED BY TALL CLIFFS, DOTTED BY NUMEROUS SEA CAVES.

231 TOP LEFT ATOKOS ISLAND IS PART OF A GROUP OF 25 SMALL ISLANDS LOCATED NEAR THE MAINLAND MOUNTAINS OF THE AKARNANIKA.

THE ISLAND OF CRETE - GREECE - THE ISLAND OF CRETE

232 TOP CRETE, THE SOUTHERNMOST GREEK ISLAND, IS CHARACTERIZED BY BEACHES WITH SOFT SAND AND TURQUOISE WATERS, HIGH MOUNTAINS WITH STEEP CLIFFS, AND A STILL PRISTINE NATURAL ENVIRONMENT.

232-233 ON CRETE, THE COASTS ARE CONSTANTLY CHANGING: SMALL BEACHES SUDDENLY GIVE WAY TO IMPERVIOUS AND SHEER SEA CLIFFS.

233 SPINALONGA PENINSULA, IN ELOUNDA BAY ON THE NORTHEAST SIDE OF CRETE, IS JOINED TO THE ISLAND BY THE NARROW ISTHMUS OF POROS.

The Island of Crete

Greece

Crete is the Mediterranean's fourth largest island; it is also the largest of the islands belonging to Greece, and it is the harshest and most challenging. Crete's natural environments and ecosystems are unique and deserve special attention. Mt. Ida, Crete's highest mountain rises over 8058 ft (2456 m), and the island is crossed by deep canyons, which give it a combination of climates and landscapes typical of mountainous territory as well as the Mediterranean. Located on Europe's southernmost latitude, Crete enjoys a mild year-round climate – and is therefore a popular spot for visitors and vacationers in the non-summer months. The landscape is varied, with mountains in the interior, highlands and deep gorges, and a coast featuring both broad beaches and areas of coves. The deep Samaria Gorge, unique of its kind in the entire Mediterranean region, together with the surrounding slopes and other smaller canyons, are highlights in what is one of the island's most interesting areas: Lefka Ori ("White Mountains") National Park.

Since the arrival of the first colonists about 8000 years ago, the island's history has featured a constant and heavy human presence, and yet the environment has remained surprisingly natural, flowery, and diverse. Crete is still renowned for the variety of its flora, its dense woods of oaks, cypresses, and chestnut trees, as well as for its rather specialized fauna. Mediterranean scrub characterizes lower altitudes, where aromatic plants predominate, while at higher elevations the landscape is essentially alpine. The island's vegetation is particularly diverse, and at least 160 species in Crete are endemic and therefore found nowhere else. Among these, for example, is one of the only two date palms native to Europe, the *Phoenix theophrasti.* The national park itself is home to more than 450 species and subspecies of the plant kingdom, among them the dittany, the ebony, and the Cretan tracheophyta pine trees.

The fauna of Crete is characterized by a remarkable degree of isolation. In fact, the island can claim two endemic mammalian species: the Cretan white-toothed shrew (*Crocidura zimmermanni*) and the Cretan spiny mouse (*Acomys minous*). Great variety is encountered among the island's invertebrates, yet among land snails and ground beetles, 40 percent are unique species. Of particular interest to naturalists is the Bezoar goat, once called "the Cretan ibex." It is actually not certain whether the animal is a "proto-goat" introduced into Crete at a time when domestication had not yet affected its appearance or whether it is an island breed. It is certain, though, that this animal has maintained a notable "wild" appearance. Difficult to spot on Crete, it is easier to observe on some of the islands close to Crete, among them Dia, Theodoros, and Ayii Pantes.

As Crete is on the main migratory route from Africa to northern Europe, the island is home to many carnivorous birds.

234 TOP LEFT THE SMALL TOWN OF CHORA SFAKION IS FAMOUS FOR ITS BEACH HIDDEN BETWEEN THE ROCKS, AT THE FOOT OF THE WHITE MOUNTAINS (LEFKA ORI).

234 TOP RIGHT THE GULF OF MIRABELLO, IN ELOUNDA BAY, IS CONSIDERED ONE OF THE MOST PRIVILEGED SPOTS IN ALL OF THE AEGEAN FROM A CLIMACTIC POINT OF VIEW.

234-235 BALOS BAY AND GRAMVOUSA TOGETHER FORM A VERY INTERESTING AREA FROM A WILDLIFE PERSPECTIVE, GIVEN THEIR GREAT ABUNDANCE OF BIRDLIFE AND THE MANY CAVES THERE THAT SOMETIMES PROVIDE REFUGE TO THE RARE MONK SEAL.

The Island of Crete

236 CRETE IS CROSSED BY MOUNTAINOUS CHAINS LIKE LEFKA ORI, MOUNT IDA, AND LASSITHI, WHOSE SILHOUETTES RUN PARALLEL TO THE SOUTHERN COASTLINE, STEEP AND CRAGGY.

236-237 SEVERAL SPRINGS ORIGINATE ON THE LASSITHI PLATEAU; THEY IRRIGATE THE ISLAND'S FERTILE PLAINS.

237 TOP IN ADDITION TO ITS MAGNIFICENT, CULTURE, LANDSCAPES, AND WILDLIFE, CRETE IS KNOWN FOR ITS SIGNIFICANT AGRICULTURAL DEVELOPMENT. THE ISLAND'S VINEYARDS, THE SOURCE OF FINE WINES, ARE AN IMPORTANT ASSET.

The Neusiedler See

Austria

The Neusiedler See, a large lake, whose basin extends across the Austrian-Hungarian border, is included in the Neusiedler See-Seewinkel National Park in the Burgenland region of easternmost Austria. The Neusiedler See is the biggest and westernmost salt lake on steppe land in Europe. About half of its surface area is occupied by a large cane-field, unique among European natural environments, which serves as a wintering site for many bird species. The lake's maximum depth is about 7 ft (2 m), but generally, it barely exceeds 3 ft (1m). The very old belt of reeds (probably dating back at least a century) that surrounds the lake extends for 4 miles (6.4 km) along the lake's western shore and about 11 miles (17 km) in the southern part. Besides the main lake, the 60 or so other small *lacken* in the Seewinkel area are important rest stops for birds.

The basin's large volume of water stabilizes temperature variations as much in summer as it does in winter. Because of its limited depth, for example, the lake heats up quickly at the end of the cold season, thus preventing any late icing over. As a result, a long growing season, lasting over 250 days, favors the growth of plants. The lake's other noteworthy special qualities include the saltiness of the water and the lack of effluents: for these reasons the Neusiedler See can be considered one of the most interesting habitats on the European continent.

The lake and *lacken*, once surrounded by deciduous forests, are now encircled by meadows, farms, and small towns. Centuries of farming have contributed to forming the lake's characteristic cultural landscape. The vegetation in the area is basically typical of the combination of salty wetlands and Eurasian steppe, with elements that reflect a mixed variety of Alpine, Baltic, and Pannonian flora. In much of the area Arctic and Alpine species grow in proximity to those from the steppes of southern Russia.

Among local amphibians are the spotted lizard, the crested lizard, the fire-bellied toad, and several frog species; among the reptiles are common lizards, sand lizards, and a few snake species. Apart from some mammals of eastern origin, like the steppe polecat and the pouched marmot, the greatest zoological interest in the area goes – even in this unique environment – to the bird life, which is very abundant and diversified, with about 300 species, half of which nest in the area. Among the most noteworthy species are the spoonbill, the white heron, the white stork, and the glossy ibis. In addition, migrating birds make significant use of vast areas of the park. Finally, the Neusiedler See-Seewinkel National Park is one of the most important habitats for a species of bird that is highly threatened worldwide: the great bustard.

238 THE NEUSIEDLER SEE IS A LAKE OF GREAT IMPORTANCE, BEING THE LARGEST SALT LAKE IN EUROPE.

238-239 AROUND THE LAKE GROW HALOPHYTIC VEGETATION AND A VAST STRIP OF CANE-FIELDS. IN SUMMER, THE NEUSIEDLER SEE EXPERIENCES DRASTIC FLUCTUATIONS IN THE WATER LEVEL.

239 TOP AROUND THE NEUSIEDLER SEE, BIRDLIFE IS VERY PLENTIFUL. FOR THIS REASON, THE SITE WAS INCLUDED AMONG THE AREAS IN THE RAMSAR CONVENTION.

The Neusiedler See

240 THE NEUSIEDLER SEE'S CANE FIELDS ARE INHABITED BY YELLOW WAGTAILS AND BITTERNS, WHICH DIFFER FROM THE OTHER HERONS BECAUSE OF THEIR SMALL SIZE AND THE DARK UPPER PORTION OF THEIR BODIES.

241 THE MARSH HABITATS OF THE NEUSIEDLER SEE ARE HOME TO A LARGE NUMBER OF BIRD SPECIES, INCLUDING THE GREAT WHITE EGRET.

242 TOP AND 242-243 TRIGLAV NATIONAL PARK FEATURES A TYPICAL MOUNTAIN ENVIRONMENT, WITH PEAKS, SMALL ALPINE LAKES, AND WATERFALLS.

243 ONE THE MOST IDYLLIC AND LEAST KNOWN AREAS IN THE ALPINE ARC IS THE TRENTA VALLEY AND THE GORGE OF THE SOCA RIVER.

244-245 THE MASSIFS THAT MAKE UP THE JULIAN ALPS REACH AN ALTITUDE OF 9364 FT (2854 M).

The National Park of Triglav

Slovenia

The first conservation proposal for the Triglav area dates back to 1908, but no real action was taken until 1924. The national park came into being in that year primarily through the efforts of the Slovenian Museological Society's Nature Protection Section, together with the local Alpine Society. Initially, a 20-year lease was stipulated for a 3400-acre (1375-hectare) area, thereafter transformed into a "protected alpine park." In 1961, the mandate was made permanent, bringing the size of the park to about 5000 acres (2025 hectares); it was enlarged yet again, following reorganization in 1981, to 324 sq. miles (839 sq. km). The park is dominated by the centrally located Mt. Triglav, after which it is named. At 9397 ft/2864 m in height, this peak is considered to be the symbol of Slovenia. Several valleys, forming a radial pattern, run down from Mt. Triglav and contribute water to the great river systems of the Soca and Sava, respectively flowing into the Adriatic and the Black Sea.

The park is home to many glacial lakes, among them the Bohinj, the Triglav Lakes, the Kriz Lakes, and the Kr Lakes. Picturesque waterfalls abound; some of the most famous are those of the Savica River ("little Sava") in the area of Bohinj (which are much loved by enthusiasts of rafting and similar sports), those of the Pericnik in the Vrata Valley, and those of Skocniki in the Beli Potok. The spectacular Krma glacial valley in the park's northern area is encircled by tall peaks (generally around 6000 ft/1839m), though the valley's lowest sections barely reach 2900 ft/885 m). The proximity of these mountains to the Balkans and the Mediterranean Sea has enabled Alpine and sub-Mediterranean thermophile (heat-loving) species to colonize and thrive side by side. Endemic species are also very numerous, from the Triglav rose (*Potentilla nitida*) to the purplish red *Primula wulferiana* and from the white *Papaver julicum ernesti-mayeri* to the Bohinj iris (*Iris cengialti volhinensis*) and the Triglav gentian. Many of these species are strictly protected to prevent their extinction.

Some animals in the Triglav area have been close to extinction: the ibex, for example, had almost completely disappeared because of a long hunting season. However, thanks to a re-introduction program, the ibex has been able to reestablish itself. Recently, another important and positive fact has been noted: some mammals have been expanding their territories and wandering into areas adjoining the park, whose northwestern corner borders on Italy, running adjacent to a tract of the Tarvisio Forest close to the Austrian border. In June 2004, a Triglav brown bear was spotted in the northern part of park, right on the border with Italy and Austria.

246 TOP AND 246-247 FLOWING WATER HAS BEEN A KEY ELEMENT IN THE FORMATION AND BEAUTY OF THE LANDSCAPES OF THE PLITVICE NATIONAL PARK.

247 THE CONTINUOUS CREATION AND DEPOSIT OF A CALCAREOUS CRUST ON AQUATIC VEGETATION FORMS CATARACTS AND WATERFALLS, AS IN THE CASE OF THE KORANA RIVER.

The Plitvice Lakes

Croatia

The magic of Plitvice has a simple name: water. Shaped by water, which appears in a veritable profusion of waterfalls, ponds, and streams, this park, a proud symbol of Croatia, is characterized by crystal-clear waters, countless small lakes, and a varied landscape of travertine stone, a calcareous rock with a warm, light brown color. All this is Plitvice, one of the most charming protected areas in Croatia and all of Europe. The national park (not to mention the UNESCO Biosphere Reserve) is located between Zagreb and Zara, in a lush green valley with an oceanic climate, wedged between forest-clad mountains. The Bijela Rijeka and Crna Rijeka rivers ("the White River" and "the Black River"), both tributaries of the Korana, make Plitvice an enchanting gem. In fact, in spreading out and transforming into hundreds of cascades, little waterfalls, and small streams, the two waterways link Plitvice's 16 lakes into a sort of dense network.

The lakes and basins have their origin in karstic springs that emerge when water flows are blocked by zones of impermeable dolomitic rock. The first of the lakes, the Prosce, is located about 1970 ft (600 m), whereas the last, the Novakovic, is located at about 1740 ft (530 m) and collects all the waters of Plitvice into a waterfall of the same name; from a height of 250 ft (76 m) it thunders down into the Korana River. Over the course of thousands of years, the waters have created underwater karstic conduits with dolines, swallow-holes, and caves. The origin of the Plitvice "natural garden" was actually a biodynamic process – the continuous formation and subsequent destruction of the travertine barriers built up from the sedimentation of carbonates. Plant life also played an important role in the formation of the Plitvice basins; mosses actively contribute to the shaping of the travertine rock, and algae encourage encrustation.

From a geological and naturalistic point of view, the park can in fact be divided into two sections. The first section – the upper one – is composed of a dolomitic-origin valley, with lakes surrounded by lush forests. In the second section – the lower one – smaller lakes and less dense vegetation appear. Then, mountains alternate with valleys blanketed by forests, rocks, marshes, and underground cavities created by karstic activity.

Plitvice has highly diverse flora that includes marsh and riparian species, as well as large forests of silver fir, spruce, and beech. Among the many mammals living in the park are brown bears, wolves, lynxes, wild cats, otters, martens, roe deer, deer, wild boars, and badgers. In conclusion, the park's management practices deserve mention; the managers have made Plitvice an excellent model for many other parks. More than 2000 employees, well laid out and maintained paths, and the effective management of visitor groups, which are accompanied and led to the main attractions, make Plitvice a true "nature industry," with a significant annual revenue.

The Plitvice Lakes

248-249 SIXTEEN LITTLE LAKES HAVE FORMED IN THE GORGES OF THE KORANA RIVER; THEY ARE FAMOUS FOR THEIR CRYSTALLINE WATERS.

249 THE UPPER LAKES IN PLITVICE HAVE BASINS FORMED BY TRIASSIC DOLOMITIC ROCK, ON WHICH EROSION CAUSED THE FORMATION OF WATERFALLS AND OTHER CASCADES. THE BARRIERS THAT SEPARATE THE VARIOUS LAKES AND PONDS ARE COMPOSED OF TRAVERTINE ROCK THAT IS CONSTANTLY ERODING INTO NEW SHAPES.

250 TOP LYNXES, NOW EXTINCT IN MUCH OF THE ALPINE CHAIN, ARE STILL NUMEROUS IN THE PLITVICE LAKES.

250 CENTER DURING WINTER, PLITVICE IS TRANSFORMED INTO AN ENCHANTED LANDSCAPE, FULLY REVEALING ITS CHARME.

250 BOTTOM ALTHOUGH PLITVICE PARK IS MAINLY VISITED IN SUMMER, IT IS THE AUTUMN COLORS AND THE MAGICAL FROZEN WATERFALLS THAT MAKE IT A UNIQUE PLACE IN THE EYES OF NATURE LOVERS.

251 THE WINTER COLD TRANSFORMS WATERFALLS AND CATARACTS INTO A PROFUSION OF JETS AND FOUNTAINS, AS IN THE CASE OF THE KORANA RIVER.

The Kornati Islands Croatia

THE KORNATI ISLANDS - CROATIA - THE KORNATI ISLAND

Among the most spectacular zones on the European coasts are the Kornati (or the "Crowned") Islands. They are an extraordinary chain of calcareous, arid, and dazzling islands in Croatia, bleached by sun and rain, composed of 140 large islands and about 50 reefs. In the whole of the eastern Adriatic, the Kornati are truly unique, true "fragments of the moon" set into a blue sea. The borders of the archipelago are Dugi Otok in the northwest, the Zirje Islands in the southeast, the Pasman Islands, Vrgada, and Murter near the mainland, and the open sea to the southwest. The archipelago is about 15 miles (24 km) in length, with a maximum breadth of 8 miles (13 km).

The name of the archipelago seems to have a double origin: first, from the fact that kings of Croatia were crowned on Kornet, the biggest island, and then from the presence of peculiar morphological characteristics such as the spectacular spires and precipices of karstic origin called "crowns."

Many of the Kornati Islands are uninhabited; they consequently retain overall a wild and pristine state, a condition from which both the natural environments and wildlife benefit. The practice of traditional livestock raising is slowly diminishing, thus new areas in which holm oaks and other trees and vegetation can grow are becoming available. With regard to land fauna, birds predominate, with abundant quantities of many species native to the Mediterranean. Among the mammals, the beech martens (*Martes foina*). However, the fauna overall is lacking in insectivores and rodents, as is usually the case on islands.

The Kornati Islands' great beauty and the wilderness that reigns on them certainly had a role in the 1980 proclamation, under which the entire archipelago became a national park. The protected area covers about 85 sq. miles (220 sq. km), of which only one percent is dry land. It is, however, the sea floor that offers the most exceptional rewards. The marine biocoenoses (communities of living creatures of different species that share the same space and compete within it for survival) and the beauty of the seabed, with underwater spires, deep caves, and abundant animal life, have made the Kornati Islands and their waters a favorite destination among scuba divers from around the world.

Finally, in addition to the archipelago's naturalistic aspects, it is also worth emphasizing its physical appearance and rather out-of-the-ordinary landscapes. The islands look like true mazes of sea and rocks, with high reefs and hills of unusual shapes, which attest to the important processes in the formation of the islands themselves. The archipelago was formed as a result of the gradual sinking of the coast following the last Ice Age; it sank about 3 ft (1 m) every thousand years. Until 2000 years ago, the present-day islands of Kornat, Katina, and Dugi Otok actually formed one single island.

The Kornati Islands are located on a fault line between the African continental plate and the European one. One of the largest faults along this front extends south from the Istrian Peninsula straight out toward the southwestern limits of the islands off northern Croatian coast, ending among the islands off the central Dalmatian coast.

This "fracture" is situated exactly in the so-called "crown of Kornati," whose rocks, perpendicular to the water surface, attest to the movement of the nearby blocks of the earth's crust: in fact, these cliffs were lifted as a result of the collision of the faults. Thereafter, the continuous processes of corrosion and erosion modified their surface, so that in some areas parts of the Kornati Islands chain sank into the depths of the sea, ending perpendicularly, in some case 300 ft (1000 m) down. The part of the national park above water is composed of a typical terrain of carbonatic rocks, which display classic karstic phenomena, among them caves, canyons, and dolines.

252-253 THE KORNATI ISLANDS ARE RENOWNED FOR THEIR WILD APPEARANCE AND THE INTENSE COLOR OF THE SEA SURROUNDING THEM.

253 TOP THE CONFORMATION OF THE KORNATI ISLANDS CHANGES RAPIDLY BETWEEN THE HIGH ROCK SEA CLIFFS AND THE BARE AND FLAT AREAS.

254-255 THE KORNATI ISLANDS OCCUPY 124 SQ. MILES (321 SQ. KM). THEY TOTAL ABOUT 150 ISLANDS AND ISLETS, BLEACHED BY THE SUN, ERODED BY THE SALTY AIR, AND WHIPPED BY THE WINDS.

The Hortobágy National Park

Hungary

THE HORTOBÁGY NATIONAL PARK - HUNGARY - THE HO

The *puszta* are the last great plains of Europe and are among the European landscapes and territories that can boast an ancient history of longstanding relationships with the local culture. In Hungarian, the word *puszta* means "deserted or abandoned." These terms clearly indicate the *puszta*'s somewhat "desolate" appearance, but in no way is this description intended to detract from the plains' poetic qualities and the richness of the local customs.

The history of the *puszta* dates back to the last Ice Age, when it began to take shape as the vast floodplains of the Tisza and Berettyò rivers. These two waterways, with their periodic floods, ensured the survival of the lakes and extensive marshlands, alternating with large areas of riparian forests (meaning those typical of the shores of waterways or lakes). Nonetheless, the rising waters, coupled with the dry, warm climate that has characterized the last thousand years, have caused repeated periods of alkalization in the region's soil, and in more recent times, increased precipitation has favored the growth of dense forests and swamp vegetation, alternating with grassy and now steppe-like zones. Then, the forested areas met their fate when the Turks, who invaded the area around 1300, started to cut them down. Deprived of their tree cover, the vast central-European plains were exposed to the bite of the wind, which transformed them into a large steppe perfectly suited to livestock raising.

The presence of domestic animals, among other factors, still constitutes a characteristic element of particular value to the region's folkloristic and cultural aspects. A close relationship with the presence of water was also longstanding: in fact, until a few centuries ago, the *puszta* was crossed by an elaborate system of navigable canals, and even today there are still many lakes. In the 19th century, a large project to drain the lands caused the canals and lakes to gradually dry up; as a result, only in scattered places is a wetter environment still found.

Today, the *puszta* is safeguarded and preserved mainly in the Hortobágy, which is also Hungary's largest national park. Founded in 1973 to protect and conserve the *puszta*'s naturalistic and cultural aspects, the park now extends across 200,000 acres (80,950 hectares). In 1999, Hortobágy was designated a UNESCO Biosphere Reserve, and a third of its area is protected by the Ramsar Convention for Wetlands. From a wildlife perspective, the *puszta* and Hortobágy Park are characterized, above all by the existing of an extremely rich and varied birdlife. A significant count of 332 observable species makes these areas true European paradises for ornithologists. The large flocks of migratory birds are noteworthy, and include cranes and marsh birds, such as the white-fronted goose (*Anser albifrons* – whose worldwide number appears to be decreasing), many other species of geese and herons, and also the great bustard.

In addition to protecting avian life, the many missions of the Hortobágy include safeguarding some significant livestock species, such as the *racka* sheep with their immense horns, the grey bulls, and the Mangalica pig. This last, which in many cases has returned to a wild or semi-wild state, is now fairly common. Overall, this trio represents very interesting ancient varieties, in many ways still primitive, whose biodiversity is certainly worth preserving. Together with the Nonius horses, these semi-domestic animals are now an integral part of the *puszta*, whose vast expanses were once a place for the herds of horses led out to graze by the *csikos* (horse-keepers). For over 2000 years the traditional use of the *puszta* as pasturelands was effectively guaranteed by a pastoral society. Since they worked with horses, the *csikos* were, in fact, the highest-placed figures in the hierarchy of animal-tenders, above the cowherds (the *gulya*) and the shepherds (the *juhasz*). The *csikos* were in many ways the true "guardians" of Hungary: they were estimated to number a few thousand up to a couple centuries ago. Today they have unfortunately been reduced to a few dozen and their centuries-old culture is being lost.

256-257 MANY ZONES IN THE PUSZTA ARE CHARACTERIZED BY CANE-FIELDS AND SWAMPS, THE REMAINS OF WHAT WERE ONCE THE VAST WETLANDS OF CENTRAL EUROPE.

257 TOP IN SPRING, BLOOMING MEADOWS COVER THE HORTOBÁGY PUSZTA, ONE OF THE LAST AREAS TO PRESERVE THE FEEL OF ANCIENT EUROPE.

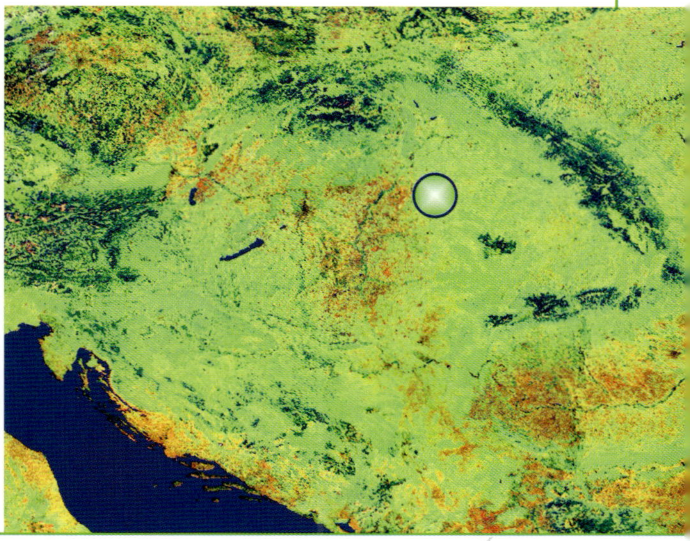

The Hortobágy

258 BIRDS ARE THE MOST IMPORTANT FAUNISTIC ELEMENT IN THE HUNGARIAN PUSZTA, AND HERONS ARE AMONG THE MOST NOTEWORTHY SPECIES.

259 TOP THE MUSKRAT IS AN AQUATIC ANIMAL RECENTLY INTRODUCED TO THE HORTOBÁG AREA.

259 CENTER AND BOTTOM PURPLE HERONS (TOP) AND THE WHITE-FRONTED GOOSE (BOTTOM) ARE SPECIES TYPICAL OF THE HORTOBÁGY. WITH OVER 200 SPECIES OF MIGRATORY BIRDS, THE PARK IS A TRUE TEMPLE FOR ORNITHOLOGISTS FROM ALL OVER EUROPE.

DANUBE RIVER DELTA - ROMANIA - DANUBE RIVER DELTA

260 TOP LEFT THE DANUBE DELTA IS ONE OF THE LAST TRUE WILD AREAS IN EUROPE

260 TOP RIGHT AND 260-261 THE DANUBE DELTA ENCOMPASSES SWAMPS, MARSHLANDS WITH ENDLESS CANE-FIELDS, EMERGENT DUNES, AND SANDY STRIPS OF FLUVIAL AND MARINE ORIGIN, NOT TO MENTION LAKES, CANALS, AND LITTLE ISLANDS. ADDITIONALLY, IT IS A NESTING PLACE FOR MANY SPECIES OF BIRDS.

Danube River Delta

Romania

Flowing into the Black Sea, after a course of 1773 miles (2853 km) and after having collected the waters of over 300 tributaries, the Danube creates a world of savage and primordial beauty in its delta, one of the few nature sanctuaries in the heart of Europe. Located on a latitude halfway between the North Pole and the Equator, the delta extends over 1600 sq. miles (4143 sq. km), of which 1330 (3444 sq. km) are in Romanian territory and the other 300 (776 sq. km) are in Ukraine.

Fifty miles before reaching the Black Sea, the Danube divides into four branches, the Chilia, Tulcea, Sulina, and Sfîntu Gheorghe, which create a landscape unique in the world. Each of these diversions actually forms a maze of large and small lakes, canals, dunes, marshes, cane fields, and willow groves. The Danube Delta constitutes the largest marshlands in Europe, in which only 10 percent of the territory is permanently above the water level; in addition, it has the valid claim of being the "youngest" area in all of Europe, in as much as the deposited terrain has been laid down only very recently.

At second glance confirms that the delta it is one of Europe's most exotic and fascinating zones, with over 1200 plant and 300 bird species. On the Old Continent, the Danube Delta represents the equivalent of the great African or Asian spaces, where nature lives day to day as the main character and, above all, where it truly shapes life patterns. With its wealth of food resources for avian life, the delta is a true "paradise for birds," representing a crossroads and place for resting or passing through for many of the migratory species that come from Egypt or the Caucasus. The passage, marked by water and the union of river and sea, winds across a series of isolated lakes, swamp zones, small rivers, brackish waters, stagnant pools, reed marshes, and dunes. Here grows what is probably the vastest cane-field on the entire planet. The canes, predominantly of the *Phragmites australis* species, form numerous little islands called *plaur*, which emerge from swamps dotted with water lilies.

The reserve includes 18 distinct zones, with protected areas and buffer areas (extending across a good 1.5 million acres (607,082 hectares), in which activities such as tree cutting and reed gathering are allowed, alongside strictly protected zones occupying a surface area of 123,000 acres (49,776 hectares) where human activities are strictly regulated. The true symbol of the delta, from an ornithological point of view, is the pelican, or better yet, the pelicans, since the largest colonies of Europe's two species, the white pelican and the Dalmatian pelican, are found here. The white pelican is in fact the most numerous species, with over 2000 pairs, whereas the Dalmatian pelican numbers only a few dozen pairs. The nesting areas are not normally visitable, and only when the little ones grow up are a few researchers allowed to approach the colonies, using small boats made especially for this purpose. Generally, viewing pelicans is easier along the Sfîntu Gheorghe Channel. From October to November, these birds leave and migrate to the Nile delta, their favorite place to winter, returning to the Danube in the spring.

Other species of particular importance are the small cormorant (with 2500 nesting pairs), numerous ibises, herons, storks, and many white swans. Finally, there are also many mammals, among them a few species (the otter, wild cats, and the muskrat) currently underrepresented in Europe. In addition, the delta is home to 150 species of fish, most notably the renowned sturgeon.

In 1990, recognizing the Danube Delta's exceptional geological and naturalistic character, not to mention its great variety of ecosystems, UNESCO declared it to be a Biosphere Reserve. In 1991, UNESCO inscribed it on the Heritage of Humanity List and in the Ramsar Convention.

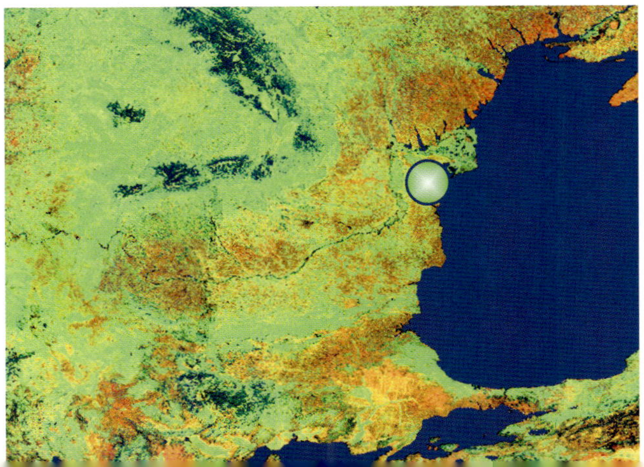

Danube River Delta

262 TOP THE WHISKERED TERN IS ONE OF THE MANY SPECIES OF MARSH AND LIMICOLOUS BIRDS TYPICAL OF THE DANUBE DELTA.

262 BOTTOM ABUNDANT FISH LIFE SUPPORTS THE PRESENCE OF NUMEROUS COLONIES OF FISH-EATING BIRDS, LIKE THE PYGMY CORMORANT.

262-263 AND 263 TOP THE POPULATION OF WHITE PELICANS HAS EXPERIENCED A REMARKABLE INCREASE IN RECENT YEARS AND REPRESENTS ONE OF THE GREATEST WILDLIFE ATTRACTIONS 'N THE DANUBE DELTA.

264 TOP AND 265 ELBRUS IS THE BEST-KNOWN PEAK IN THE CAUCASUS, WHICH EXTENDS FROM THE BLACK SEA TO THE CASPIAN.

264-265 DOMBAY-ULGEN REACHES AN ALTITUDE OF 13,278 FT (4046 M) AND IS ONE OF THE MOST FASCINATING PEAKS IN THE WESTERN CAUCASUS, WHICH REPRESENTS A TRUE GEOGRAPHICAL BARRIER.

266-267 THE BEZENGI MASSIF IS ALSO KNOWN AS THE "HIMALAYAS IN MINIATURE," AS IT ENCOMPASSES THE TALLEST PEAKS IN THE CAUCASUS.

The Caucasus

The Russian Federation

Extending over an area of 680,000 acres (275,185 hectares), the western section of the Caucasus is one of the most important mountain systems in Europe (though some geographers would place the range in Eurasia), with peaks that range from just under 6000 ft (1828 m) to the 13,278 ft (4046 m) of Mt. Dombay-Ulgen. Between the Caucasus and Lesser Caucasus highlands with their continental climate is a flat, densely populated area divided into the two basins by the Surami Mountains. This range links the Great Caucasus to the north with the Lesser Caucasus to the south. To the west of the Surami Mountains, the basin of the Rioni River looks onto the Black Sea, and to the east, the vast valley of the Kura River extends through Georgia toward Azerbaijan and the Caspian Sea. The geology of the area includes formations of igneous, metamorphic, and sedimentary rock, though in the northern part, it is dominated by calcareous massifs on which erosive action that occurred during the Ice Age gouged out enormous cracks. These happen to contain the largest cave in Russia, 9 miles (14.5 km) long and 5250 ft (1600 m) deep. The range's southern slopes have many rivers, which flow into the Black Sea.

The Western Caucasus is one of the few mountain areas in Europe that has not have been subject to excessively damaging human impact. Many wild animals populate its pasturelands. Overall, the Caucasus encompasses a vast diversity of ecosystems; also it is also the place of origin (and now reintroduction) of the European bison. There is good reason that the area has been classified among the "hot spots" of world biodiversity; both the large number of animal and plant species and the region's role as a true crossroads between different zoological-geographical regions justify the term. In fact, more than 10,000 plants, 700 vertebrates, and 20,000 invertebrates, have been cataloged here. Among the 60 species of mammals that maintain stable populations inside the Caucasus Reserve and the Sochi National Park are wolves, bears, wild boar, Caucasian deer, lynxes, chamois, and the Kuban River basin kar (*Capra caucasica*), a high-mountain goat. Vegetation extends over several altitude zones. Up to about 3900 ft (1200 m) there are deciduous forests in which oaks, chestnut trees, horn-beams, eastern beeches, and pear trees predominate, giving way to coniferous forest at higher altitudes.

c = caption
bold = devoted chapter

INDEX - INDEX - INDEX - INDEX - INDEX - INDEX - INDEX

Page 1 left Marcello Bertinetti/Archivio White Star
Page 1 center Antonio Attini/Archivio White Star
Page 1 right Marcello Libra/Archivio White Star
Pages 2-3 Hannu Hautala
Pages 4-5 Olivier Grunewald
Page 6 Motor-Presse Syndicat/Agefotostock/Marka
Page 7 Giovanni Simeone/Sime/Sie
Pages 8-9 Jorma Lutha/Naturepl/Contrasto
Pages 10-11 Rainer Hackenberg/zefa/Corbis
Page 13 Marcello Bertinetti/Archivio White Star
Page 15 Sven Zellner
Pages 16-17 Mario Colonel
Pages 18-19 Antonio Attini/Archivio White Star
Pages 20-21 Laurent Nedelec
Page 22 Yann Arthus-Bertrand/Corbis
Page 56 top and bottom Giulio Veggi/Archivio White Star
Pages 56-57 Giulio Veggi/Archivio White Star
Page 58 Eide Per/Corbis Sygma/Corbis
Pages 58-59 Wofgang Chonen/Agefotostock/Marka
Page 59 Wofgang Chonen/Agefotostock/Marka
Page 60 left Maurizio Biancarelli
Page 60 right Peter Cairns
Pages 60-61 Peter Cairns
Page 61 bottom Maurizio Biancarelli
Pages 62-63 Peter Cairns
Page 63 top Maurizio Biancarelli
Page 63 bottom Marcello Libra
Page 64 right William Paton/Nhpa
Pages 64-65 Mark Hamblin
Page 103 right Maurizio Biancarelli
Pages 104-105 Jean AND. Roche/Naturepl/Contrasto
Page 105 top Nigel Bean/Naturepl/Contrasto
Page 105 bottom Stuart Franklin/Magnum Photos/Contrasto
Pages 106-107 Carsten Braun
Pages 108-109 Alamy Images
Page 109 left N. Thibaut/Hoa-Qui/Hachette Photo/Contrasto
Pages 110-111 Alamy Images
Page 112 right Vincent Munier/Naturepl/Contrasto
Pages 112-113 Vincent Munier/Naturepl/Contrasto
Page 113 Vincent Munier/Naturepl/Contrasto
Pages 114-115 Vincent Munier/Naturepl/Contrasto
Page 115 top and bottom Vincent Munier

PHOTOGRAPHIC CREDITS - PHOTOGRAPHIC CREDITS - PH

Pages 22-23 Yann Arthus-Bertrand/Corbis
Page 23 right Ragnar Th. Sigurdsson/Arctic Images
Page 24 top and bottom Yann Arthus-Bertrand/Corbis
Pages 24-25 Yann Arthus-Bertrand/Corbis
Pages 26-27 Hans Strand
Page 28 top and left Ragnar Th. Sigurdsson/Arctic Images
Pages 28-29 Ragnar Th. Sigurdsson/Arctic Images
Page 29 Ragnar Th. Sigurdsson/Arctic Images
Pages 30-31 Ragnar Th. Sigurdsson/Arctic Images
Page 32 left Giulio Veggi/Archivio White Star
Pages 32-33 G. Popp & V. Hackner
Page 33 Kjell Ljungström/Naturbild
Page 34 top G. Popp & V. Hackner
Page 34 bottom Vitantonio Dell'Orto
Pages 35 Bengt Hedberg/Naturbild
Page 37 Vitantonio Dell'Orto
Pages 36-37 Kjell Ljungström/Naturbild/Naturepl/Contrasto
Page 37 top Marcello Libra
Page 37 center B & C Alexander/Nhpa
Page 37 bottom Vitantonio Dell'Orto
Pages 38-39 Jorma Lutha/Naturepl/Contrasto
Page 39 top and bottom Jorma Lutha/Naturepl/Contrasto
Page 40 Riccardo Spila/Sime/Sie
Pages 40-41 Piergirgio Scalandris/Sime/Sie
Page 41 left J-F Hellio & N. Van Ingen
Page 42 Riccardo Spila/Sime/Sie
Pages 42-43 Hubert Stadler/Corbis
Page 43 top, center and bottom Marcello Libra/Archivio White Star
Pages 44-45 Marcello Libra/Archivio White Star
Page 45 top and bottom Marcello Libra/Archivio White Star
Page 46 Hannu Hautala
Pages 46-47 Hannu Hautala
Page 47 right Hannu Hautala
Pages 48-49 Hannu Hautala
Pages 50-51 Hannu Hautala
Page 51 top and bottom Hannu Hautala
Pages 52-53 Hannu Hautala/Panda Photo
Page 54 left Chris Liesle/Corbis
Page 54 right R. Segato/Panda Photo
Pages 54-55 Hasse Schröder/Naturbild
Page 55 bottom Vitantonio Dell'Orto
Page 65 Pete Cairns/Naturepl/Contrasto
Page 66 Mark Hamblin
Pages 66-67 Peter Cairns
Page 67 top Andy Rouse/Nhpa
Page 67 bottom Peter Cairns
Pages 68-69 Peter Cairns
Page 70 right John Downer/Naturepl/Contrasto
Pages 70-71 Massimo Ripani/Sime/Sie
Page 71 Adrian Warren/Lastrefuge.co.uk
Pages 72-73 Dae Sasitorn/Lastrefuge.co.uk
Page 74 top Jon Arnold/DanitaDelimont
Page 74 bottom Julie Royle/Nhpa
Pages 74-75 Adrian Warren/Lastrefuge.co.uk
Pages 76-77 Adrian Warren/Lastrefuge.co.uk
Page 78 left and right Antonio Attini/Archivio White Star
Pages 78-79 Antonio Attini/Archivio White Star
Page 79 bottom Alamy Images
Page 80 Antonio Attini/Archivio White Star
Pages 80-81 Antonio Attini/Archivio White Star
Page 82 right Bernard Castelein/Naturepl/Contrasto
Pages 82-83 Vitantonio Dell'Orto
Page 83 Milko Marchetti
Page 84 left right Renato Fano and Anna Flagiello
Pages 84-85 Bernard Castelein/Naturepl/Contrasto
Page 85 Renato Fano and Anna Flagiello
Page 86 right M. Lanini/Panda Photo
Pages 86-87 Sven Zellner
Page 87 Sven Zellner
Page 88 Sven Zellner
Page 89 top Desiree Astrom
Page 89 bottom Sven Zellner
Page 90 Eric Baccega/Naturepl/Contrasto
Pages 90-91 Sven Zellner
Page 91 top Gunter Grafenhain/Sime/Sie
Page 91 bottom Maurizio Biancarelli
Pages 92-93 Sven Zellner
Page 94 right Marcello Libra/Archivio White Star
Pages 94-95 K. Haenel/Corbis
Page 95 Marcello Libra/Archivio White Star
Pages 96-97 Ingo Arndt
Page 98 Marcello Libra/Archivio White Star
Pages 98-99 Marcello Libra/Archivio White Star
Page 99 right Marcello Libra/Archivio White Star
Pages 100-101 Cornelia Dorr
Page 102 Raymond Gerhman/Corbis
Pages 102-103 Niall Benvie/Naturepl/Contrasto
Pages 116-117 Vincent Munier
Page 118 left Jean AND. Roche/Naturepl/Contrasto
Pages 118-119 Kevin O'Hara/Agefotostock/Marka
Page 119 top and bottom Jean AND. Roche/Naturepl/Contrasto
Pages 120-121 Frank Lukasseck/Agefotostock/Marka
Page 122 left J-F Hellio & N. Van Ingen
Page 122 right top Guido Alberto Rossi/Tips Images
Page 122 center J-F Hellio & N. Van Ingen
Page 122 right bottom J-F Hellio & N. Van Ingen
Page 123 J-F Hellio & N. Van Ingen
Page 124 J-F Hellio & N. Van Ingen
Pages 124-125 Christopher Ratier/Nhpa
Page 125 top Maurizio Biancarelli/Panda Photo
Page 125 bottom Photononstop/Photolibrary
Page 126 right Christian Vaisse/Hoa-Qui/Hachette Photo/Contrasto
Pages 126-127 Chris Liesle/Corbis
Page 127 Antonio Attini/Archivio White Star
Pages 128-129 Antonio Attini/Archivio White Star
Page 130 Marcello Bertinetti/Archivio White Star
Pages 130-131 N. Thibaut/Hoa-Qui/Hachette Photo/Contrasto
Page 131 left Marcello Bertinetti/Archivio White Star
Pages 132-133 Marcello Bertinetti/Archivio White Star
Page 134 right Carsten Braun
Pages 134-135 A. Nardi/Panda Photo
Page 135 top M. Piacentino/Panda Photo
Page 135 bottom J-F Hellio & N. Van Ingen
Page 136 J-F Hellio & N. Van Ingen
Pagg 136-137 Alberto Nardi/Nhpa
Page 137 top Maurizio Biancarelli
Page 137 bottom Thomas Dressler
Page 138 left Laurent Nedelec
Page 138 right Jordi Bas
Pages 138-139 Jordi Bas
Page 140 Jordi Bas
Pages 140-141 Laurent Nedelec
Page 141 top and bottom Jordi Bas
Pages 142-143 Jordi Bas
Page 144 Laurent Nedelec
Pages 144-145 Jordi Bas
Page 145 left Jordi Bas
Page 145 right Jose Luis Gomez de Franciso/Naturepl/Contrasto
Page 146 left and right Giulio Veggi/Archivio White Star

Pages 146-147 Andres Blomqvist/Lonely Planet Images
Page 147 center Giulio Veggi/Archivio White Star
Page 147 bottom Alberto Nardi/Nhpa
Page 148 Giulio Veggi/Archivio White Star
Pages 148-149 Giulio Veggi/Archivio White Star
Page 149 left and right Giulio Veggi/Archivio White Star
Page 150 right Greg Gawlowski/Lonely Planet Images
Pages 150-151 Mario Colonel
Page 151 Marco Milani/K3
Pages 152-153 Siegfried Eigstler/Sime/Sie
Page 154 left and right Ingo Arndt
Pages 154-155 Marcello Libra/Archivio White Star
Page 155 Marcello Libra/Archivio White Star
Page 182 right Franco Cogoli/Sime/Sie
Pages 182-183 Johanna Huber/Sime/Sie
Page 184 left and right Antonio Attini/Archivio White Star
Pages 184-185 G. Marcoaldi/Panda Photo
Page 185 bottom Antonio Attini/Archivio White Star
Page 186 Milko Marchetti
Pages 186-187 Milko Marchetti
Page 187 left, center and right Milko Marchetti
Page 188 left Maurizio Biancarelli
Page 188 right Vitantonio Dell'Orto
Pages 188-189 L. Longo/Panda Photo
Pages 190-191 Maurizio Biancarelli
Page 192 left Maurizio Biancarelli
Page 192 right Giulio Veggi/Archivio White Star
Pages 228-229 Massimo Ripani/Sime/Sie
Page 230 Johanna Huber/Sime/Sie
Pages 230-231 Yann Arthus-Bertrand/Corbis
Page 231 left Michael St. Maur Sheil/Corbis
Page 231 center Giovanni Simeone/Sime/Sie
Page 231 right Massimo Ripani/Sime/Sie
Page 232 right Marco Simoni/Agefotostock/Marka
Pages 232-233 Philippos Katsiyiannis/www.greeknature.com
Page 233 Giovanni Simeone/Sime/Sie
Page 234 left Johanna Huber/Sime/Sie
Page 234 right Giovanni Simeone/Sime/Sie
Pages 234-235 Johanna Huber/Sime/Sie
Page 236 Doug Pearson/Sime/Sie
Pages 236-237 Jon Arnold/DanitaDelimont

TOGRAPHIC CREDITS - PHOTOGRAPHIC CREDITS - PHOTO-

Pages 156-157 Mario Colonel
Page 158 right Marcello Bertinetti/Archivio White Star
Pages 158-159 Mario Colonel
Page 159 Marcello Bertinetti/Archivio White Star
Pages 160-161 Mario Colonel
Page 161 top and bottom A. Nardi/Panda Photo
Page 162 left Mario Colonel
Page 162 right Marcello Bertinetti/Archivio White Star
Pages 162-163 Marcello Bertinetti/Archivio White Star
Page 163 Mario Colonel
Pages 164-165 Marcello Bertinetti/Archivio White Star
Page 165 Marco Milani/K3
Pages 166-167 Mario Colonel
Page 168 Alessandro Gogna/K3
Pages 168-169 G. Pollini/Panda Photo
Page 169 left Marco Milani/K3
Page 170 left G. Marcoaldi/Panda Photo
Page 170 right Didier Givois/Agefotostock/Marka
Pages 170-171 Marco Milani/K3
Page 171 top Alessandro Gogna/K3
Page 171 bottom Alamy Images
Pages 172-173 Luciano Gaudenzio/K3
Page 173 top Luciano Gaudenzio/K3
Page 173 center M. B ranchi/Panda Photo
Page 173 bottom Carsten Braun
Pages 174-175 Marco Milani/K3
Page 175 top Stefano Unterthiner
Page 175 center John Waters/Naturepl/Contrasto
Page 175 bottom Alejandro Diaz Diez/Agefotostock/Marka
Page 176 bottom Marcello Bertinetti/Archivio White Star
Pages 176-177 Marcello Bertinetti/Archivio White Star
Page 177 left and right Marcello Bertinetti/Archivio White Star
Pages 178-179 Marcello Bertinetti/Archivio White Star
Page 180 top and center Marcello Bertinetti/Archivio White Star
Page 180 bottom Giulio Veggi/Archivio White Star
Pages 180-181 Marcello Bertinetti/Archivio White Star
Page 182 left Olimpo Fantuz/Sime/Sie
Pages 192-193 Maurizio Biancarelli
Page 193 Maurizio Biancarelli
Pages 194-195 Giulio Veggi/Archivio White Star
Page 196 left Marcello Bertinetti/Archivio White Star
Pages 196-197 Marcello Bertinetti/Archivio White Star
Page 197 Antonio Attini/Archivio White Star
Pages 198-199 Marcello Bertinetti/Archivio White Star
Page 200 Marcello Bertinetti/Archivio White Star
Page 201 top, center and bottom Marcello Bertinetti/Archivio White Star
Pages 202-203 Marcello Bertinetti/Archivio White Star
Page 204 Antonio Attini/Archivio White Star
Pages 204-205 Giulio Veggi/Archivio White Star
Page 205 right Antonio Attini/Archivio White Star
Page 206 Antonio Attini/Archivio White Star
Pages 206-207 Giulio Veggi/Archivio White Star
Page 207 top Antonio Attini/Archivio White Star
Page 207 bottom Giulio Veggi/Archivio White Star
Page 208 Giulio Veggi/Archivio White Star
Pages 208-209 Giovanni Simeone/Sime/Sie
Page 210 Antonio Attini/Archivio White Star
Pages 210-211 Antonio Attini/Archivio White Star
Page 211 right Olivier Grunewald
Page 212 top and bottom Antonio Attini/Archivio White Star
Pages 212-213 Antonio Attini/Archivio White Star
Page 213 Antonio Attini/Archivio White Star
Page 214 Alessandro Saffo/Sime/Sie
Page 215 top Alessandro Saffo/Sime/Sie
Page 215 bottom Giulio Veggi/Archivio White Star
Pages 216-217 Alessandro Saffo/Sime/Sie
Page 218 right Reinhard Schmid/Sime/Sie
Pages 218-219 Reinhard Schmid/Sime/Sie
Page 219 Reinhard Schmid/Sime/Sie
Pages 220-221 MarkaCollection
Page 222 Philippos Katsiyiannis/www.greeknature.com
Pages 222-223 L. Vinco/Panda Photo
Page 223 left L. Vinco/Panda Photo
Pages 224-225 Philippos Katsiyiannis/www.greeknature.com
Page 226 bottom Alamy Images
Pages 226-227 Giovanni Simeone/Sime/Sie
Page 227 left Giovanni Simeone/Sime/Sie
Page 227 right Massimo Ripani/Sime/Sie
Page 228 Massimo Ripani/Sime/Sie
Page 237 left Doug Pearson/Sime/Sie
Page 237 right Kevin Schafer
Page 238 top and bottom G. Popp & V. Hackner
Pages 238-239 G. Popp & V. Hackner
Page 239 right Renato Fano and Anna Flagiello
Page 240 top Renato Fano and Anna Flagiello
Page 240 bottom Marcello Libra
Page 241 Renato Fano and Anna Flagiello
Page 242 right G. Popp & V. Hackner
Pages 242-243 G. Popp & V. Hackner
Page 243 Armando Maniciati
Pages 244-245 Marco Milani/K3
Page 246 right Marcello Libra/Archivio White Star
Pages 246-247 Giovanni Simeone/Sime/Sie
Page 247 Maurizio Biancarelli
Pages 248-249 Giovanni Simeone/Sime/Sie
Page 249 top and bottom Marcello Libra/Archivio White Star
Page 250 top, center and bottom Maurizio Biancarelli
Page 251 top and bottom Maurizio Biancarelli
Pages 252-253 Jonathan Blair/Corbis
Page 253 left Arne Hodalic/Corbis
Page 253 right Alamy Images
Pages 254-255 Jonathan Blair/Corbis
Pages 256-257 Gunter Ziesler/Agefotostock/Marka
Page 256 left Massimo Crivellari
Page 258 Bence Maté
Page 259 top Renato Fano and Anna Flagiello
Page 259 center and bottom Renato Fano and Anna Flagiello
Page 260 left Marcello Libra
Page 260 right Mike Potts/Naturepl/Contrasto
Pages 260-261 Marcello Libra
Page 262 top and bottom Marcello Libra
Pages 262-263 Marcello Libra
Page 263 Ed Kashi/Corbis
Page 264 right Georgi G. Shablovsky/Lonely Planet Images
Pages 264-265 Konstantin Mikhailov/Naturepl/Contrasto
Page 265 Dean Conger/Corbis
Pages 266-267 Konstantin Mikhailov/Naturepl/Contrasto
Page 272 Antonio Attini/Archivio White Star

All maps are of BNSC 1998/NERC/RAL/CCLRC/ESA

272 A LOW-LYING AND SANDY SHORE DESCENDS INTO THE WATERS OF THE ATLANTIC IN NORTHWEST IRELAND: THIS IS VAST DONEGAL BAY, INTO WHICH EMPTY THE WATERS OF THREE SHORT RIVERS FED BY GLACIAL LAKES LOCATED INLAND. THESE ROUGH-EDGED BASINS ARE THE LAST EXISTING EVIDENCE OF A TIME WHEN THE ISLAND WAS COVERED BY VAST GLACIERS THAT SCRAPED AWAY THE SURFACE OF THE LAND.